Studies in Systems, Decision and Control

Volume 159

Series editor

Janusz Kacprzyk, Polish Academy of Sciences, Warsaw, Poland
e-mail: kacprzyk@ibspan.waw.pl

The series "Studies in Systems, Decision and Control" (SSDC) covers both new developments and advances, as well as the state of the art, in the various areas of broadly perceived systems, decision making and control–quickly, up to date and with a high quality. The intent is to cover the theory, applications, and perspectives on the state of the art and future developments relevant to systems, decision making, control, complex processes and related areas, as embedded in the fields of engineering, computer science, physics, economics, social and life sciences, as well as the paradigms and methodologies behind them. The series contains monographs, textbooks, lecture notes and edited volumes in systems, decision making and control spanning the areas of Cyber-Physical Systems, Autonomous Systems, Sensor Networks, Control Systems, Energy Systems, Automotive Systems, Biological Systems, Vehicular Networking and Connected Vehicles, Aerospace Systems, Automation, Manufacturing, Smart Grids, Nonlinear Systems, Power Systems, Robotics, Social Systems, Economic Systems and other. Of particular value to both the contributors and the readership are the short publication timeframe and the world-wide distribution and exposure which enable both a wide and rapid dissemination of research output.

More information about this series at http://www.springer.com/series/13304

Andrew Schumann · Krzysztof Pancerz

High-Level Models of Unconventional Computations

A Case of Plasmodium

Springer

Andrew Schumann
University of Information Technology
  and Management in Rzeszow
Rzeszów
Poland

Krzysztof Pancerz
Department of Computer Science,
  Faculty of Mathematics and Natural
  Sciences
University of Rzeszów
Rzeszów
Poland

ISSN 2198-4182   ISSN 2198-4190 (electronic)
Studies in Systems, Decision and Control
ISBN 978-3-030-06295-8   ISBN 978-3-319-91773-3 (eBook)
https://doi.org/10.1007/978-3-319-91773-3

Printed on acid-free paper

This Springer imprint is published by the registered company Springer International Publishing AG part of Springer Nature
The registered company address is: Gewerbestrasse 11, 6330 Cham, Switzerland

Contents

Chapter 1
Introduction

Physarum polycephalum, called also *slime mould*, belongs to the species of order *Physarales*, subclass *Myxogastromycetidae*, class *Myxomycetes*, division *Myxostelida*. Plasmodium is its vegetative phase represented as a single cell with a myriad of diploid nuclei. The Physarum Chips, designed in our project *Physarum Chip Project: Growing Computers From Slime Mould* supported by the Seventh Framework Programme (FP7-ICT2011-8),[1] are programmed by spatio-temporal configurations of repelling and attracting gradients. About this computer, called the *slime mould computer*, please see [4–6, 8, 19, 31, 33, 34, 41, 42, 57, 97, 102, 104, 124, 134, 143, 146]. This computer is an organic extension of *reaction-diffusion computer*, about the latter please see [7, 10, 32, 107]. The idea of *protein robots* as computers designed on actin filament networks is an extension of slime mould computers, in turn, see [11, 14, 75, 101, 105, 134, 138, 140, 141]. There are several classes of Physarum Chips: morphological processors, sensing devices, frequency-based, bio-molecular and microfluidic logical circuits, and electronic devices. These Chips are based on actin filament networks: [11, 14, 101, 105, 138–142].

The *Physarum polycephalum* plasmodium behaves and moves as a giant amoeba. Typically, the plasmodium forms a network of protoplasmic tubes connecting the masses of protoplasm at the food sources which has been shown to be efficient in terms of network length and resilience [4]. In the project we have proposed high-level programming tools for the Physarum Chips in the form of a new object-oriented programming language [60, 62, 116, 117, 119, 126]. Within this language we can check possibilities of practical implementations of storage modification machines on plasmodia and their applications to behavioural science such as behavioural economics and game theory. The proposed language can be used for developing programs for the slime mould by the spatial configuration of stationary nodes.

---

[1]For more details please see http://www.phychip.eu/.

A. Schumann and K. Pancerz, *High-Level Models of Unconventional Computations*, Studies in Systems, Decision and Control 159, https://doi.org/10.1007/978-3-319-91773-3_1

The plasmodium can be interpreted as transition system $\mathscr{S} = (States, Edges)$, where (i) $States$ is a set of states presented by attractants occupying by the plasmodium, (ii) $Edges \subseteq States \times States$ is the set of transitions presenting the plasmodium propagation. Transitions can be defined as logic gates within different logical systems (classical as well as non-classical). So, we can deal formally with different transition systems depending on ways how we define transitions, by means of which logics.

Theoretically, transition systems are studied within coinductive calculus of streams [73, 74] and coalgebras [72]. Behavioural equivalence in transition systems is understood as bisimulation [35, 39, 71]. The arithmetic operations of coinductive calculus of streams are the same as the arithmetic operations on p-adic integers [83]. The difference of coinductive calculus of streams and coalgebras from p-adic analysis is that while the first branches were created within computer science as theoretical framework for transition systems, the latter was developed on the basis of topology with the non-Archimedean property [16, 50, 52]. In other words, if we want to study topology of streams, we should appeal to p-adic analysis. This analysis can be used for quantum mechanics [47], biological modelling [45], and coinductive probability theory [48, 49, 80, 82, 83].

We face transition systems everywhere in intelligent behaviour, e.g. in business processes [130, 131]. While coinductive calculus of streams and coalgebras are their theoretical framework, in programming they are reconstructed within object-oriented programming languages. The point is that this kind of programming defines not only the data type of a data structure, but also the types of operations (functions) that can be applied to the data structure. Therefore the data structure becomes an object that includes both data and functions. It is impossible to program real transition systems like business processes in another way. Thus, the object-oriented programming language is a high-level computer programming language that implements objects and their associated procedures within the programming context to create software programs [28]. The concepts used in the object-oriented programming are formalized in coinductive calculus of streams and coalgebras.

In our work, we use coinductive calculus of streams, coalgebras, and p-adic analysis as theoretical frameworks for reconstructing *Physarum* transition systems [125]. For coding their behaviour, we appeal to object-oriented programming, where we should start with defining objects (both data and functions of a data structure). It can be done differently, e.g. by means of different logics in defining transitions, by means of different properties of attractants and active zones of plasmodium, etc.

In programming slime mould, first of all, we have constructed logic gates through the proper geometrical distribution of stimuli. This approach has been adopted from the ladder diagram language [129] widely used to program *Programmable Logic Controllers* (PLCs). Flowing power has been replaced with propagation of plasmodium of *Physarum polycephalum*. Plasmodium propagation is stimulated by attractants and repellents and rungs of the ladder can consist of serial or parallel connected paths of *Physarum* propagation. A kind of connection depends on the arrangement of regions of influences of individual stimuli. If both stimuli influence *Physarum*, we obtain alternative paths for its propagation. It corresponds to a parallel connection (i.e. the OR gate). If the stimuli influence *Physarum* sequentially, at the beginning only the first one, then the second one, we obtain a serial connection (i.e. the AND gate). The NOT gate is imitated by the repellent avoiding *Physarum* propagation.

In the proposed approach, we assume that each attractant (repellent) is characterized by its region of influence in the form of a circle surrounding the location point of the attractant (repellent), i.e. its center point. The intensity determining the force of attracting (repelling) decreases as the distance from it increases. A radius of the circle can be set assuming some threshold value of the force. The plasmodium must occur in a proper region to be influenced by a given stimulus. This region is determined by the radius depending on the intensity of the stimulus. Controlling the plasmodium propagation is realised by activating/deactivating stimuli.

Logic values for inputs have the following meaning in terms of states of stimuli: 0—attractant/repellent deactivated, 1—attractant/repellent activated. Logic values for outputs have the following meaning in terms of states of stimuli: 0—absence of *Physarum polycephalum* at the attractant, 1—presence of *Physarum polycephalum* at the attractant.

Then we have adopted more abstract models than distribution of stimuli to program *Physarum polycephalum* machines which can be identified with programming in the high-level language. At the beginning the choice fell on Petri nets, first developed by C. A. Petri, see [13, 67–69, 119]. They are a powerful graphical language for describing processes in digital hardware. We have shown how to build Petri net models, and next implement *Physarum polycephalum* machines by using basic logic gates AND, OR, NOT, and their simple combination circuits. In our approach, we use Petri nets with inhibitor arcs. Inhibitor arcs are used to disable transitions, they test the absence of tokens at a place. A transition can proceed only if all its places connected through inhibitor arcs are empty. This ability of Petri nets with inhibitor arcs is used to model behaviour of repellents. Plasmodium of *Physarum* avoids light and some thermo- and salt-based conditions and this fact can be modelled by inhibitor arcs. The Petri net model (code in the high-level language) can be translated into the code in the low-level language, i.e. geometrical distribution of attractants and repellents of the *Physarum* machine.

In the object-oriented programming language for simulating the plasmodium motions we are based on process-algebraic formalizations of *Physarum* storage modification machine [108, 112]. So, we consider some instructions in *Physarum* machines in the terms of process algebra like as follows [54]: add node, remove node, add edge, remove edge. Adding and removing nodes can be implemented through activation and deactivation of attractants, respectively. Adding and removing edges can be implemented by means of repellents put in proper places in the space. An activated repellent can avoid a plasmodium transition between attractants. Adding and removing edges can change dynamically over time. To model such a behaviour, we propose a high-level model, based on timed transition systems. In this model we define the following four basic forms of *Physarum* transitions (motions): *direct* (direction, a movement from one point, where the plasmodium is located, towards another point, where there is a neighbouring attractant), *fuse* (fusion of two plasmodia at the point, where they meet the same attractant), *split* (splitting plasmodium from one active point into two active points, where two neighbouring attractants with a similar power of intensity are located), and *repel* (repelling of plasmodium or inaction).

In *Physarum* motions, we can perceive some ambiguity influencing on exact anticipation of states of *Physarum* machines in time. In case of splitting plasmodium, there is some uncertainty in determining next active points (attractants occupied by plasmodium) if a given active point is known. This uncertainty does not occur in case of direction, where the next active point is uniquely determined. To model ambiguity in anticipation of states of *Physarum* machines, we propose to use rough set theory. Analogously to the lower and upper approximations, we define the lower and upper predecessor anticipations of states in the *Physarum* machine. The behaviour of *Physarum* machines can also be modelled using Bayesian networks with probabilities defined on rough sets [61].

Thus, we propose some timed and probabilistic extensions of standard process algebra to implement timed and rough set models of behaviour of *Physarum* machines in our new object-oriented programming language, called by us the *Physarum language*, for *Physarum polycephalum* computing. In this language we can program the slime mould.

In this book, we show that the plasmodium propagation is a natural process algebra (labelled transition system). So, basing on it, we can propose high-level programming models for controlling the slime mould behaviour. Our programming of *Physarum polycephalum* is a pure behaviorism: we consider possibilities of simulating all basic stimulus-reaction relations. Slime mould is a good experimental medium for behavioristic models. Hence, the programming tools for modelling slime mould, proposed in the book, can be applied in different behavioral sciences which are based on studying the stimulus-reaction relations.

In Chap. 2, we regard the slime mould propagation as a labelled transition system. In Chap. 3, we define *Physarum* machines as such. Then we define Petri net models for them (Chap. 4). Further, we concentrate on rough set extensions of plasmodium transition systems (Chap. 5). Then we define the notion of non-well-foundedness (Chap. 6) and the *Physarum* language (Chap. 7). The next chapter (Chap. 8) is devoted to p-adic valued logic, where $p - 1$ is the number of possible attractants. In Chap. 9, we offer p-adic valued arithmetic gates in plasmodium transitions. In Chap. 10, we define bio-inspired games as a high-level model of slime mould transitions. In Chap. 11, we consider Go games—the games in the 5-adic valued universe. In Chap. 12, we propose game-theoretic interfaces for simulating slime mould.

Chapter 2
Natural Labelled Transition Systems and Physarum Spatial Logic

Usually, a *labelled transition system* is used for describing the behaviour and tempo-spatial structure of concurrent systems [148]. The latter were first introduced by Milner [54]. Since that time many logics for concurrent systems have been built up [29, 37] including logics aimed to describe spatial properties of mobile processes [23–26]. Luis Caires and Luca Cardelli introduced *spatial logics* [23, 24], which are able to specify systems that deal with fresh or secret resources such as keys, nonces, channels, and locations.

It is worth noting that the interactive-computing paradigm proposed by Milner can describe concurrent (parallel) computations whose configuration may change during the computation and is decentralized as well. Within the framework of this paradigm, one proposed a lot of so-called *concurrency calculi* also called *process algebras*. They are typically presented using systems of equations. These formalisms for concurrent systems are formal in the sense that they represent systems by expressions and then reason about systems by manipulating the corresponding expressions. The behaviour of plasmodium of *Physarum polycephalum* shows an instance of one of the natural implementations of concurrent systems. Thus, the plasmodium should be considered as a parallel computing substrate. It is one of the natural examples of concurrent and mobile computational processes as such.

The plasmodium forms characteristic veins of protoplasm in looking for the food sources and it is very intelligent in building transporting networks [41, 135–137, 144, 145]. It is light-sensitive, which gives us additional means to program its motions. *Physarum* exhibits articulated negative phototaxis. Therefore by using masks of illumination one can control dynamics of localizations in these media: change a signal's trajectory or even stop a signal's propagation, amplify the signal, generate trains of signals, etc. [4, 41, 136, 137, 143, 146, 147].

The main reason to consider the behaviour of plasmodium of *Physarum poly-cephalum* within Milner's paradigm of concurrent computation is that this behaviour

© Springer International Publishing AG, part of Springer Nature 2019
A. Schumann and K. Pancerz, *High-Level Models of Unconventional Computations*, Studies in Systems, Decision and Control 159,
https://doi.org/10.1007/978-3-319-91773-3_2

could serve as a natural implementation of labelled transition system and spatial logic. In analyzing the plasmodium we observe processes of fusion and choice that could be interpreted as unconventional (spatial) conjunction and disjunction denoted by & and + respectively. Both operations differ from conventional ones, because they cannot have a denotational semantics in the standard way. However, they may be described within spatial logic. This shows that many (if not all) natural systems like *Physarum polycephalum* should be regarded beyond the set-theoretic axiom of foundation [46], i.e. beyond the *von Neumann's sequential paradigm of computation*, but at the same time they may be examined within the *Milner's interactive-computing paradigm*.

2.1 Experimental Data

All the experiments for us were performed by Andrew Adamatzky's team at the University of the West of England, Bristol. According to their experiences, the plasmodia of *Physarum polycephalum* (slime mould) were cultured on wet paper towels, fed with oat flakes, and moistened regularly. They subcultured the plasmodium every 5–7 days. Experiments were performed in standard Petri dishes, 9 cm in diameter. Depending on particular experiments they used 2% agar gel or moisten filter paper, nutrient-poor substrates, and 2% oatmeal agar, nutrient-rich substrate (Sigma-Aldrich). All experiments were conducted in a room with diffusive light of 3–5 cd/m, 22 °C temperature. In each experiment an oat flake colonized by the plasmodium was placed on a substrate in a Petri dish, and few intact oat flakes distributed on the substrate. The intact oat flakes acted as source of nutrients, attractants for the plasmodium. Petri dishes with plasmodium were scanned on a standard HP scanner. The only editing done to scanned images is color enhancement: increase of saturation and contrast.

Repellents were implemented with illumination domains using blue electroluminescent sheets, see details in [4]. Masks were prepared from black plastic, namely the triangle was cut in the plastic, when this mask was placed on top of the electroluminescent sheet, the light was passing only through the cuts.

2.2 Physarum Process Calculus

Assume that the computational domain S is partitioned into computational cells s_j, $j = 1, \ldots, K$ such that $s_i \cap s_j = \emptyset$, $i \neq j$ and $\bigcup_{j=1}^{K} s_j = S$. Each computational cell contains just one activated or deactivated attractant or repellent.

Further, suppose that there are $N < K$ active species or growing pseudopodia and the state of cells i is denoted by s_i, $i = 1, \ldots, K$. These states are time dependent and they change by occupation and deoccupation by the plasmodium through the activation and deactivation of attractants or repellents. Hence, plasmodium's active zones interact concurrently in this way.

Foraging the plasmodium can be represented as a set of the following abstract entities [30]:

1. The set of active zones (growing pseudopodia) $S_{init} = \{s_1, s_2, \ldots\}$, i.e., the set of initial states in the plasmodium propagation. On a nutrient-rich substrate plasmodium propagates as a typical circular, target, wave, while on the nutrient-poor substrates protoplasmic tubes or pseudopodia are formed.
2. The set of attractants $A = \{e_1, e_2, \ldots\}$ are sources of nutrients, on which the plasmodium feeds. It is still subject of discussion how exactly the plasmodium feels presence of attractants. But experimentally we see that the plasmodium can locate and colonize nearby sources of nutrients. Each attractant e_i is a function propagating the plasmodium from one state s_k to another state s_m. It is possible that the plasmodium does not propagate its pseudopodia (when the neighbour attractants are deactivated), so its transition is nil in this case.
3. The set of repellents $R = \{e'_1, e'_2, \ldots\}$. The plasmodium of *Physarum poly-cephalum* avoids light. Thus, domains of high illumination are repellents such that each repellent e' is characterized by its position and intensity of illumination, or force of repelling. In other words, each repellent e' is a function from one state s_k to another state s_m, too.
4. The set of protoplasmic tubes or pseudopodia $T = S \times E \times S$, where $E = A \cup R$. Typically plasmodium spans sources of nutrients with protoplasmic tubes/veins. The plasmodium builds a planar graph, where nodes are sources of nutrients, e.g., oat flakes, and edges are protoplasmic tubes. They can be considered transitions from S to S.

Basing on these entities, we can consider the slime mould propagation as a natural transition system. Abstract transition systems are a commonly used and understood model of computation. A transition system consists of a set of states, with an initial state, together with transitions between states. Transitions are labelled to specify the kind of events they represent (cf. [148]). Labelled transitions systems were originally introduced as named transition systems in [44]. In general, we can consider transition systems with a set of initial states instead of a single initial state.

Thus, we adopt the following definition of a transition system [108, 111, 112].

Definition 2.1 A transition system is a quadruple $TS = (S, E, T, S_{init})$, where:

- S is the non-empty set of states,
- E is the set of events,
- $T \subseteq S \times E \times S$ is the transition relation,
- $S_{init} \subseteq S$ is the set of initial states.

Let $E = \{e_1, e_2, \ldots, e'_1, e'_2, \ldots\}$ be a set of names consisting of attractants and repellents. With every $e \in E$ we associate a complementary action $\bar{e}$. Define $L = \{e, \bar{e} : e \in E\}$, where e is considered as activator (an activation of attractant or repellent) and $\bar{e}$ as inhibitor for e (a deactivation of an appropriate attractant or repellent), being the set of labels built on E (under this interpretation, $e = \bar{\bar{e}}$). Suppose that an event e communicates with its complement $\bar{e}$ to produce the internal action τ. Define $L_\tau = L \cup \{\tau\}$.

Definition 2.2 A labelled transition system is a quadruple $TS_L = (S, L, T, S_{init})$, where:

- S is the non-empty set of states,
- L is the set of labels,
- $T \subseteq S \times L \times S$ is the transition relation,
- $S_{init} \subseteq S$ is the set of initial states.

The sets E and L of both definitions are considered actions which may be viewed as labeled events. If $(s, e, s') \in T$, then the idea is that TS can go from s to s' as a result of the event e occurring at s. A single element $(s, e, s') \in T$ is called shortly a transition. We can write a transition as

$$s \xrightarrow{e} s'.$$

This notation corresponds to a graphical representation of transition systems (see Fig. 2.1).

Any transition system $TS = (S, E, T, S_{init})$ or $TS_L = (S, L, T, S_{init})$ can be presented in the form of a labeled directed graph with nodes corresponding to states from S, edges representing the transition relation T, and labels of edges corresponding to events from E or L. Initial states are encircled to distinguish them.

Example 2.1 Let us consider a transition system $TS = (S, E, T, S_{init})$, where:

- $S = \{s_1, s_2, s_3, s_4, s_5\}$,
- $E = \{e_1, e_2, e_3, e_4\}$,
- $T = \{(s_1, e_1, s_2), (s_1, e_2, s_3), (s_1, e_3, s_4), (s_2, e_4, s_5)\}$,
- $S_{init} = \{s_1\}$.

The transition systemn TS can be presented in the form of a labeled directed graph shown in Fig. 2.1. The initial state s_1 is encircled.

It is sometimes convenient to consider transition between states as strings of events. We write

$$s_1 \xrightarrow{v} s_k,$$

Fig. 2.1 The transition system TS presented in the form of a labeled directed graph

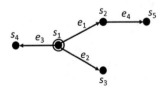

where $v = e_1 e_2 \ldots e_{k-1}$ is a, possibly empty, string of some events from E, to mean

$$s_1 \xrightarrow{e_1} s_2 \xrightarrow{e_2} \ldots \xrightarrow{e_{k-1}} s_k$$

for some states $s_1, s_2, \ldots, s_k$ from S.

Example 2.2 Let us consider a transition system $TS = (S, E, T, S_{init})$ given in Example 2.1. For instance, there exists a non-empty string $v = e_1 e_4$, where $e_1, e_4 \in E$, such that $s_1 \xrightarrow{v} s_5$.

We use the symbols α, β, etc., to range over labels of L_τ, with $\alpha = \overline{\overline{\alpha}}$, and the symbols P, Q, etc., to range over processes on states s_i, $i = 1, \ldots, K$.

Our *process calculus* contains the following basic operators: 'Nil' (inaction), '*' (prefix), '|' (cooperation), '\' (hiding), '&' (reaction/fusion), '+' (choice), a (constant or restriction to a stable state), $A(\cdot)$ (attraction), $R(\cdot)$ (repelling), $C(\cdot)$ (spreading/diffusion).

Definition 2.3 The processes of TS and TS_L are given by the syntax:

$$P, Q :: = \text{Nil} \mid \alpha * P \mid A(\alpha) * P \mid R(\alpha) * P \mid C(\alpha) \mid (P|Q) \mid P\backslash Q \mid P\&Q \mid P + Q \mid a$$

Each label is a process, but not vice versa. An operational semantics for this syntax is defined as follows:

Prefix: $$\dfrac{}{\alpha * P \xrightarrow{\alpha} P}, \qquad \dfrac{}{A(\alpha) * P \xrightarrow{\beta} P}(A(\alpha) = \beta),$$

$$\dfrac{}{R(\alpha) * P \xrightarrow{\beta} P}(R(\alpha) = \beta),$$

(the conclusion states that the process of the form $\alpha * P$ (resp. $A(\alpha) * P$ or $R(\alpha) * P$) may engage in α (resp. $A(\alpha)$ or $R(\alpha)$) and thereafter they behave like P; in the presentations of behaviours as trees, $\alpha * P$ (resp. $A(\alpha) * P$ or $R(\alpha) * P$) is understood as an edge with two nodes: α (resp. $A(\alpha)$ or $R(\alpha)$) and the first action of P),

Diffusion: $$\dfrac{P \xrightarrow{\alpha} P'}{P \xrightarrow{\alpha} C(\alpha)} \qquad (C(\alpha) ::= P'),$$

Constant: $$\dfrac{P \xrightarrow{\alpha} P'}{a \xrightarrow{\alpha} P'} \qquad (a ::= P, a \in L_\tau),$$

Choice: $$\dfrac{P \xrightarrow{\alpha} P'}{P + Q \xrightarrow{\alpha} P'}, \qquad \dfrac{Q \xrightarrow{\alpha} Q'}{P + Q \xrightarrow{\alpha} Q'}$$

(these both rules state that a system of the form $P + Q$ saves the transitions of its subsystems P and Q),

$$\text{Cooperation:} \quad \frac{P \xrightarrow{\alpha} P'}{P|Q \xrightarrow{\alpha} P'|Q}, \quad \frac{Q \xrightarrow{\alpha} Q'}{P|Q \xrightarrow{\alpha} P|Q'}$$

(according to these rules, the cooperation | interleaves the transitions of its subsystems),

$$\frac{P \xrightarrow{\alpha} P' \quad Q \xrightarrow{\overline{\alpha}} Q'}{P|Q \xrightarrow{\tau} P'|Q'}$$

(i.e. subsystems may synchronize in the internal action τ on complementary actions α and $\overline{\alpha}$),

$$\text{Hiding:} \quad \frac{P \xrightarrow{\alpha} P'}{P \backslash Q \xrightarrow{\alpha} P' \backslash Q} \quad (\alpha \notin Q, \ Q \subseteq L),$$

(this rule allows actions not mentioned in Q to be performed by $P \backslash Q$),

$$\text{Fusion:} \quad \frac{}{\alpha * P \& \overline{P} \xrightarrow{\alpha} \text{Nil}}$$

(the fusion of complementary processes are to be performed into the inaction),

$$\frac{P \xrightarrow{\alpha} P' \quad Q \xrightarrow{\alpha} P'}{P \& Q \xrightarrow{\alpha} P'}, \quad \frac{P \xrightarrow{\alpha} P' \quad Q \xrightarrow{\alpha} P'}{Q \& P \xrightarrow{\alpha} P'}$$

(this means that if we obtain the same result P' that is produced by the same action α and evaluates from two different processes P and Q, then P' may be obtained by that action α started from the fusion $P \& Q$ or $Q \& P$),

$$\frac{P \xrightarrow{\alpha} P'}{P \& Q \xrightarrow{\alpha} \text{Nil} + C(\alpha) + P'}, \quad \frac{P \xrightarrow{\alpha} P'}{Q \& P \xrightarrow{\alpha} \text{Nil} + C(\alpha) + P'}$$

(these rules state that if the result P' is produced by the action α from the processes P, then a fusion $P \& Q$ (or $Q \& P$) is transformed by that same α either into the inaction or diffusion or process P').

These are inference rules for basic operations. The ternary relation $P \xrightarrow{\alpha} P'$ means that the initial action P is capable of engaging in action α and then behaving like P'.

The informal meanings of basic operations are as follows:

1. Nil, this is the empty process which does nothing. In other words, Nil represents the component which is not capable of performing any activities: a deadlocked component.

2. $\alpha * P$, a process $\alpha \in L$ followed by the process P: P becomes active only after
   the action α has been performed. An activator $\alpha \in L$ followed by the process
   P is interpreted as branching pseudopodia into two or more pseudopodia, when
   the site of branching represents newly formed process $\alpha * P$.
   In turn, an inhibitor $\overline{\alpha} \in L$ followed by the process P is annihilating protoplasmic
   strands forming a process at their intersection.

3. $A(\alpha) * P$ denotes a process that waits for a value α and then continues as P.
   This means that an attractor A modifies propagation vector of action α towards
   P. Attractants are sources of nutrients. When such a source is colonized by
   plasmodium the nutrients are exhausted and attracts ceases to function: $A(\alpha) *$
   Nil.

4. $R(\alpha) * P$ denotes a process that waits for a value α and then continues as P.
   This means that a repellent R modifies propagation vector of action α towards P.
   Process can be cancelled, or annihilated, by a repellent: $R(\alpha) * $ Nil. This happens
   when propagating localized pseudopodium α enters the domain of repellent, e.g.
   illuminated domain, and α does not have a chance to divert or split.

5. $C(\alpha)$, a diffusion of activator $\alpha \in L$ is observed in placing sources of nutrients
   nearby the protoplasmic tubes belonging to α or inactive zone ($\alpha ::= $ Nil). More
   precisely, diffusion generates propagating processes which establish a protoplasm vein (the case of activator α) or annihilate it (when source of nutrients
   exhausted, the case of inhibitor $\overline{\alpha}$).

6. $P|Q$, this is a parallel composition (commutative and associative) of actions: P
   and Q are performed in parallel. The parallel composition may appear in the
   case, two more food sources are added to either side of the array and then the
   plasmodium sends two streams outwards to engulf the sources. When the food
   sources have been engulfed, the plasmodium shifts in position by redistributing
   its component parts to cover the area created by the addition of the two new
   processes P and Q that will already behave in parallel.
   Process P can be split, or multiplied, by two sources of attractants $(A_1(A_2(P)) *$
   $P_1|P_2$. Pseudopodium P approaches the site where distance to A_1 is the same as
   distance to A_2. Then P subdivides itself onto two pseudopodia P_1 and P_2. Each
   of the pseudopodia travels to its unique source of attractants. Also, process P can
   be split, or multiplied, by a repellent: $R(P) * P_1|P_2$. The fission happens when
   a propagating pseudopodium 'hits' a repellent. The part of pseudopodium most
   affected by the repellent ceases propagating, while two distant parts continue
   their development. Thus, two separate pseudopodia are formed.

7. $P\backslash Q$, this restriction operator allows us to force some of P's actions not to occur;
   all of the actions in the set $Q \subseteq L$ are prohibited, i.e. the component $P\backslash Q$ behaves
   as P except that any activities of types within the set Q are hidden, meaning that
   their type is not visible outside the component upon completion.

8. $P\&Q$, this is the fusion of P and Q; $P\&Q$ represents a system which may behave
   as both component P and Q. For instance, Nil behaves as $P\&\overline{P}$, where P is an
   activator and $\overline{P}$ an appropriate inhibitor respectively. The fusion of P and Q is
   understood as collision of two active zones P and Q. When they collide they
   fuse and annihilate, $P\&Q* $ Nil. Depending on the particular circumstances the

new active zone α (the result of fusing) may become inactive (Nil), transform to protoplasmic tubes ($C(\alpha)$), or remain active and continue propagation in a new direction (the case of prefix $*$).

9. $P + Q$, this is the choice between P and Q; $P + Q$ represents a system which may behave either as component P or as Q. Thus the first activity to complete identifies one of the components which is selected as the component that continues to evolve; the other component is discarded. In *Physarum* calculi, the choice $P + Q$ between processes P and Q sometimes is represented by competition between pseudopodia tubes. In other words, two processes P and Q can compete with each, during this competition one process 'pulls' protoplasm from another process, thus making this another process inactive. The competition happens via protoplasmic tube.

10. a, constants belonging to labels are components whose meaning is given by equations such as $a ::= P$. Here the constant a is given the behaviour of the component P. Constants can be used to describe infinite behaviours, via mutually recursive defining equations.

2.3 Spatial Logic of Physarum Process Calculus

Now let us construct a *spatial logic of Physarum process calculus*.

Definition 2.4 Given a set L of names, an infinite set $V = \{x, y, z, \ldots\}$ of name variables, and an infinite set $\mathbf{P} = \{A, B, C, \ldots\}$ of propositional variables (mutually disjoint from the set L of names), formulas of spatial logic are defined as follows

$$\Phi ::= \text{Nil} \mid 0 \mid 1 \mid x * A \mid A(x) * A \mid R(x) * A \mid C(x) \mid (A|B) \mid A\backslash B \mid A\&B \mid A + B \mid x \mid A$$

where 0 is a falsehood constant, 1 a truth constant and other operations ($x * A$, $A(x) * A$, $R(x) * A$, $C(x)$, $A|B$, $A\backslash B$, $A\&B$, $A + B$) are the same as in the previous section.

Some derived connectives are as follows:

$$\neg A ::= 1\backslash A \quad \text{(negation)},$$
$$A \wedge B ::= A\backslash(1\backslash B) \quad \text{(conjunction)},$$
$$A \vee B ::= 1\backslash((1\backslash A)\backslash B) \quad \text{(disjunction)},$$
$$A \supset B ::= 1\backslash(A\backslash B) \quad \text{(implication)}.$$

Further, let us define a substitution s:

1. If V' is a finite set of name variables, and L is any set of names, a substitution s is a mapping assigning $s(x) \in L$ to each $x \in V'$, and x to each $x \notin V'$ (thus, outside its domain, any substitution behaves like the identity).

2. For any formula Φ and substitution s we denote by $s(\Phi)$ the formula inductively
defined as follows:

$$s(\text{Nil}) ::= \text{Nil},$$
$$s(0) \text{ is undefined},$$
$$s(1) \text{ is undefined},$$
$$s(C(x)) ::= C(s(x)),$$
$$s(x * A) ::= s(x) * s(A),$$
$$s(A(x) * A) ::= A(s(x)) * s(A),$$
$$s(R(x) * A) ::= R(s(x)) * s(A),$$
$$s(A|B) ::= s(A)|s(B),$$
$$s(A \backslash B) ::= s(A) \backslash s(B),$$
$$s(A \& B) ::= s(A) \& s(B),$$
$$s(A + B) ::= s(A) + s(B).$$

Now define a congruence relation $\cong$ on the set of processes. Assume that this
relation satisfies the following requirements for all processes:

$$P \cong P,$$
$$P \cong Q \supset Q \cong P,$$
$$(P \cong Q \wedge Q \cong R) \supset P \cong R,$$
$$P \cong Q \supset \alpha * P \cong \alpha * Q,$$
$$P \cong Q \supset A(\alpha) * P \cong A(\alpha) * Q,$$
$$P \cong Q \supset R(\alpha) * P \cong R(\alpha) * Q,$$
$$P \cong Q \supset P|R \cong Q|R,$$
$$P \cong Q \supset R|P \cong R|Q,$$
$$P \cong Q \supset P \& R \cong Q \& R,$$
$$P \cong Q \supset R \& P \cong R \& Q,$$
$$P \cong Q \supset P + R \cong Q + R,$$
$$P \cong Q \supset R + P \cong R + Q,$$

Inaction:

$$P|\text{Nil} \cong P,$$
$$\alpha * \text{Nil} \cong \text{Nil},$$

Cooperation:

$$P|Q \cong Q|P,$$
$$P|(Q|R) \cong (P|Q)|R,$$
$$P|(Q + R) \cong (P|Q) + (P|R),$$
$$P|(Q\&R) \cong (P|Q)\&(P|R),$$

Hiding:

$$\text{Nil}\backslash P \cong \text{Nil},$$
$$(P + Q)\backslash P' \cong P\backslash P' + Q\backslash P',$$
$$(P\&Q)\backslash P' \cong P\backslash P'\&Q\backslash P',$$

Fusion:

$$P\&\overline{P} \cong \text{Nil},$$
$$P\&P \cong P,$$
$$P\&\text{Nil} \cong \text{Nil},$$
$$P\&Q \cong Q\&P,$$
$$P\&(Q\&R) \cong (P\&Q)\&R,$$

Choice:

$$P + P \cong P,$$
$$P + Nil \cong P,$$
$$P + Q \cong Q + P,$$
$$P + (Q + R) \cong (P + Q) + R,$$
$$P\&(Q + R) \cong (P\&Q) + (P\&R),$$
$$P + (Q\&R) \cong (P + Q)\&(P + R).$$

The *property set* (*Pset*) of a formula Φ is a set of processes which satisfy Φ, this set is closed under $\cong$ and has a finite support (for details see [23–26]):

1. if a process P satisfies some property Φ, then any process Q such that $P \cong Q$ must also satisfy Φ;
2. if there exists a finite set of names L' such that, for all $n, m \notin L'$, P satisfies some property Φ, then $P\{n \leftrightarrow m\}$ must also satisfy Φ, where $P\{n \leftrightarrow m\}$ is a result of name transposition in P, i.e. the substitution that assigns m to n and n to m.

The semantics of formulas is defined by assigning to each formula Φ a Pset $[\Phi]_v$. A set $[\Phi]_v$ is said to be the *denotation* of a formula Φ with respect to a valuation v that assigns to each propositional variable of Φ an appropriate Pset. On the other hand, every process P belonging to a Pset has a characteristic formula $||P||$. Let $||Nil|||:: = Nil$, $||P\#Q||:: = ||P||\#||Q||$, where $\# \in \{|, \&, +, \backslash\}$; $||\alpha * P||:: = ||\alpha|| * ||P||$, $||A(\alpha) * P||:: = A(||\alpha||) * ||P||$, $||R(\alpha) * P||:: = R(||\alpha||) * ||P||$, $||C(\alpha)|| = C(||\alpha||)$, $||\alpha|| \in L$. The formula $||P||$ identifies P up to structural equivalence: for all P and Q, Q satisfies $||P||$ if and only if $Q \cong P$.

A *valuation* v is a mapping from the set of formulas assigning to each formula a set $[\Phi]_v$ of processes such that

1. for any name variable x, $v(x)$ is a result of substitution s,
2. $[C(x)]_v:: = \{P: P \cong C(s(x))\}$,
3. $[0]_v:: = \emptyset$,
4. $[1]_v:: = \Sigma$, where Σ is a universe of all processes,
5. $[Nil]_v:: = \{P: P \cong Nil\}$, i.e. the formula Nil is satisfied by any process in the structural congruence class of Nil,
6. $[\Phi|\Psi]_v:: = [\Phi]_v|[\Psi]_v = \{P: \exists Q, R.P \cong Q|R \text{ and } Q \in [\Phi]_v \wedge R \in [\Psi]_v\}$,
7. $[\Phi\&\Psi]_v:: = [\Phi]_v\&[\Psi]_v = \{P: \exists Q, R.P \cong Q\&R \text{ and } Q \in [\Phi]_v \wedge R \in [\Psi]_v\}$,
8. $[\Phi + \Psi]_v:: = [\Phi]_v + [\Psi]_v = \{P: \exists Q, R.P \cong Q + R \text{ and } Q \in [\Phi]_v \wedge R \in [\Psi]_v\}$,
9. $[\Phi\backslash\Psi]_v:: = [\Phi]_v\backslash[\Psi]_v = \{P: \exists Q, R.P \cong Q\backslash R \text{ and } Q \in [\Phi]_v \wedge R \in [\Psi]_v\}$,
10. $[\Phi \wedge \Psi]_v:: = [\Phi]_v \cap [\Psi]_v = \{P: \exists Q, R.P \cong Q \cap R \text{ and } Q \in [\Phi]_v \wedge R \in [\Psi]_v\}$,
11. $[\Phi \vee \Psi]_v:: = [\Phi]_v \cup [\Psi]_v = \{P: \exists Q, R.P \cong Q \cup R \text{ and } Q \in [\Phi]_v \wedge R \in [\Psi]_v\}$,
12. $[\neg\Phi]_v:: = \Sigma\backslash[\Phi]_v$,
13. $[x * \Phi]_v:: = \{P: P \cong (s(x) * [\Phi]_v)\}$,
14. $[A(x) * \Phi]_v:: = \{P: P \cong (s(A(x)) * [\Phi]_v)\}$,
15. $[R(x) * \Phi]_v:: = \{P: P \cong (s(R(x)) * [\Phi]_v)\}$.

We write $P \models_v \Phi$ whenever $P \in [\Phi]_v$: this means that P satisfies formula Φ under valuation v. We say that formula Φ is *coinductively valid* if $[\Phi]_v \neq \emptyset$ for any valuation v and Φ is *inductively valid* if $[\Phi]_v = \Sigma$ for any valuation v. The coinductive validity predicate is denoted by $\textbf{vld}_C(\Phi)$ and the inductive validity predicate by $\textbf{vld}_I(\Phi)$. Evidently, $\textbf{vld}_I(\Phi) \subset \textbf{vld}_C(\Phi)$, i.e. $\textbf{vld}_C(\Phi)$ is weaker than $\textbf{vld}_I(\Phi)$.

Further, let us define the coinductive realizability $\textbf{real}_C(\Phi)$ as $\textbf{real}_C(\Phi):: = \neg\textbf{vld}_C(\neg\Phi)$ and inductive realizability $\textbf{real}_I(\Phi)$ as $\textbf{real}_I(\Phi):: = \neg\textbf{vld}_I(\neg\Phi)$. Obviously, $\textbf{vld}_C(\Phi) \subset \textbf{real}_C(\Phi)$, $\textbf{vld}_I(\Phi) \subset \textbf{real}_I(\Phi)$, and $\textbf{real}_I(\Phi) \subset \textbf{real}_C(\Phi)$.

Let us denote by U the collection of all Psets. We can prove the following basic propositions:

Proposition 2.1 *For all formulas Φ, Ψ and valuations v, if $||P|| = \Phi$, $||Q|| = \Psi$ and $P \cong Q$, then $[\Phi]_v = [\Psi]_v$.*

Proof This statement directly follows from the congruence properties and properties of the valuation v. □

Proposition 2.2 *Let U be a family of all Psets (i.e., each Pset of U of the form $[\Phi]_v$ satisfies the property denoted by an appropriate formula Φ), then the triple $(U; |, Nil)$ is a commutative monoid.*

Proof We should show that (i) $[\Phi]_v|[\text{Nil}]_v = [\Phi]_v$; (ii) $[\Psi]_v|[\Phi]_v = [\Phi]_v|[\Psi]_v$; (iii) $[\Psi]_v|([\Phi]_v|[\Theta]_v) = ([\Psi]_v|[\Phi]_v)|[\Theta]_v$. All these equalities may be obtained immediately, e.g. $[\Psi]_v|[\Phi]_v = \{P: \exists Q, R.P \cong Q|R \text{ and } Q \in [\Phi]_v \wedge R \in [\Psi]_v\} = \{P: \exists Q, R.P \cong R|Q \text{ and } R \in [\Psi]_v \wedge Q \in [\Phi]_v\} = [\Phi]_v|[\Psi]_v$ due to the congruence properties. $\qquad\square$

Proposition 2.3 *The quadruple* $(U; \&, +, \text{Nil})$ *is a lattice, where Nil is a minimal element, but it is not a Boolean algebra (it has no maximal element).*

Proof We could define the ordering relation $\leq$ in the quadruple $(U; \&, +, \text{Nil})$ in the following way:

1. for any $[\Psi]_v, [\Phi]_v \in U, [\Psi]_v \leq [\Phi]_v$ iff $[\Psi]_v \& [\Phi]_v = [\Psi]_v$;
2. for any $[\Psi]_v, [\Phi]_v \in U, [\Psi]_v \leq [\Phi]_v$ iff $[\Psi]_v + [\Phi]_v = [\Phi]_v$;
3. for any $[\Psi]_v, [\Phi]_v \in U, [\Psi]_v \& [\Phi]_v \leq [\Psi]_v + [\Phi]_v$.

This relation defined above is partial. In this ordered structure we have the minimal element Nil, but there is no maximal element. $\qquad\square$

This proposition shows that the inductive validity predicate, $\mathbf{vld_I}(\Phi)$ (or $\mathbf{real_I}(\Phi)$), cannot be defined on a formula Φ if Φ is a superposition just of $\&, +, $ Nil. We could define only $\mathbf{vld_C}(\Phi)$ (resp. $\mathbf{real_C}(\Phi)$) on such Φ.

Proposition 2.4 *The quadruple* $(U; \setminus, \emptyset, \Sigma)$ *is a Boolean algebra.*

Proof Let us define the ordering relation $\leq$ as follows: for any $[\Psi]_v, [\Phi]_v \in U$, $[\Psi]_v \leq [\Phi]_v$ iff $([\Phi]_v \setminus ([1]_v \setminus [\Psi]_v)) = [\Psi]_v$. In this ordered structure there is a supremum

$$\sup\{[\Psi]_v, [\Phi]_v\} = [1]_v \setminus (([1]_v \setminus [\Psi]_v) \setminus [\Phi]_v),$$

an infimum

$$\inf\{[\Psi]_v, [\Phi]_v\} = [\Psi]_v \setminus ([1]_v \setminus [\Phi]_v),$$

a minimal member $[0]_v = \emptyset$ and a maximal member $[1]_v = \Sigma$. $\qquad\square$

The latter proposition means that $\mathbf{vld_I}(\Phi)$ (resp. $\mathbf{real_I}(\Phi)$) is well defined on a formula Φ if Φ is a superposition of $\setminus$ and 1.

Now let us try to build up a sequent calculus for *Physarum* spatial logic.

A *context*, Γ or Δ, is a finite multiset of entries of the form $P : \Phi$ where P is a process and Φ is a formula. A *sequent* is a statement $\Gamma \Rightarrow \Delta$ where Γ and Δ are contexts. Suppose $\Gamma = (P_1 : \Phi_1, \ldots, P_n : \Phi_n)$ and $\Delta = (P_1 : \Psi_1, \ldots, P_m : \Psi_m)$. Then the valuation of the sequent $\Gamma \Rightarrow \Delta$, denoted by $[\Gamma \Rightarrow \Delta]_v$, is defined as follows: $(P_1 = \Phi_1 \wedge \cdots \wedge P_n = \Phi_n) \supset (P_1 = \Psi_1 \vee \cdots \vee P_m = \Psi_m)$.

Lemma 2.1 *The following statements hold true:*

1. $\mathbf{vld_I}(\Gamma, P : 0 \Rightarrow \Delta)$; $\mathbf{vld_I}(\Gamma \Rightarrow P : 1, \Delta)$;
2. $\mathbf{vld_I}(\Gamma \Rightarrow P : 0, \Delta)$ *iff* $\mathbf{vld_I}(\Gamma \Rightarrow \Delta)$; $\mathbf{vld_I}(\Gamma, P : 1 \Rightarrow \Delta)$ *iff* $\mathbf{vld_I}(\Gamma \Rightarrow \Delta)$;

3. $\mathbf{real_I}(\Gamma, P : \Phi \backslash \Psi \Rightarrow \Delta)$ *iff* $\mathbf{real_I}(\Gamma, P : \Phi \Rightarrow P : \Psi, \Delta)$;

4. $\mathbf{real_I}(\Gamma \Rightarrow P : \Phi \backslash \Psi, \Delta)$ *iff* $\mathbf{real_I}(\Gamma \Rightarrow P : \Phi, \Delta) \wedge \mathbf{real_I}(\Gamma, P : \Psi \Rightarrow \Delta)$;

5. $\mathbf{real_C}(\Gamma, P : \Phi \& \Psi \Rightarrow \Delta)$ *iff* $\forall Q, R.P \cong Q \& R \supset \mathbf{real_C}(\Gamma, Q : \Phi, R : \Psi \Rightarrow \Delta)$;

6. $\mathbf{real_C}(\Gamma \Rightarrow P : \Phi \& \Psi, \Delta)$ *iff* $\exists Q, R.P \cong Q \& R \wedge \mathbf{real_C}(\Gamma \Rightarrow Q : \Phi, \Delta) \wedge$ $\mathbf{real_C}(\Gamma \Rightarrow R : \Psi, \Delta)$;

7. $\mathbf{real_C}(\Gamma \Rightarrow P : \Phi + \Psi, \Delta)$ *iff* $\exists Q, R.P \cong Q + R \wedge \mathbf{real_C}(\Gamma \Rightarrow Q : \Phi, R : \Psi, \Delta)$;

8. $\mathbf{real_C}(\Gamma, P : \Phi + \Psi \Rightarrow \Delta)$ *iff* $\forall Q, R.P \cong Q + R \supset (\mathbf{real_C}(\Gamma, P : \Phi \Rightarrow \Delta) \wedge \mathbf{real_C}(\Gamma, P : \Psi \Rightarrow \Delta))$;

9. $\mathbf{real_C}(\Gamma \Rightarrow P : \mathrm{Nil}, \Delta)$ *iff* $P \not\cong \mathrm{Nil} \supset \mathbf{real_C}(\Gamma \Rightarrow \Delta)$;

10. $\mathbf{real_C}(\Gamma, P : \mathrm{Nil} \Rightarrow \Delta)$ *iff* $P \cong \mathrm{Nil} \supset \mathbf{real_C}(\Gamma \Rightarrow \Delta)$;

11. $\mathbf{real_C}(\Gamma \Rightarrow P : \mathrm{Nil}, \Delta)$ *iff* $P \cong Q \& \overline{Q} \supset \mathbf{real_C}(\Gamma \Rightarrow P : \Phi \& \Psi, \Delta)$;

12. $\mathbf{real_C}(\Gamma, P : \mathrm{Nil} \Rightarrow \Delta)$ *iff* $P \cong Q \& \overline{Q} \supset \mathbf{real_C}(\Gamma, P : \Phi \& \Psi \Rightarrow \Delta)$;

13. $\mathbf{real_C}(\Gamma, P : \Phi | \Psi \Rightarrow \Delta)$ *iff* $\forall Q, R.P \cong Q | R \supset \mathbf{real_C}(\Gamma, Q : \Phi, R : \Psi \Rightarrow \Delta)$;

14. $\mathbf{real_C}(\Gamma \Rightarrow P : \Phi | \Psi, \Delta)$ *iff* $\exists Q, R.P \cong Q | R \wedge \mathbf{real_C}(\Gamma \Rightarrow Q : \Phi, \Delta) \wedge$ $\mathbf{real_C}(\Gamma \Rightarrow R : \Psi, \Delta)$;

15. $\mathbf{real_C}(\Gamma, P : x * \Phi \Rightarrow \Delta)$ *iff* $\forall Q.P \cong \alpha * Q \supset \mathbf{real_C}(\Gamma, Q : \Phi \Rightarrow \Delta)$;

16. $\mathbf{real_C}(\Gamma \Rightarrow P : x * \Phi, \Delta)$ *iff* $(\forall P'.P \not\cong \alpha * P' \wedge \mathbf{real_C}(\Gamma \Rightarrow \Delta)) \vee (\exists P'.P \cong \alpha * P' \wedge \mathbf{real_C}(\Gamma \Rightarrow P' : \Psi, \Delta))$.

Proof For superpositions of $\&, +, \mathrm{Nil}$, we associate the sequent relation $\Rightarrow$ with the ordering relation of Proposition 2.3, for other superpositions this one is associated with the conventional ordering relation of Proposition 2.4. Some cases are considered in [25]. $\qquad\qquad\qquad\qquad\qquad\qquad\qquad\qquad\qquad\qquad\qquad\qquad\qquad\quad\square$

Inference rules of *Physarum* spatial logic:

$$\frac{}{\Gamma, P : 0 \Rightarrow \Delta}, \qquad \frac{\Gamma \Rightarrow \Delta}{\Gamma \Rightarrow P : 0, \Delta}, \qquad \frac{}{\Gamma \Rightarrow P : 1, \Delta},$$

$$\frac{\Gamma \Rightarrow \Delta}{\Gamma, P : 1 \Rightarrow \Delta}, \qquad \frac{P \cong Q}{\Gamma, P : \Phi \Rightarrow Q : \Phi, \Delta},$$

$$\frac{\Gamma \Rightarrow P : \Phi, \Delta \qquad \Gamma, P : \Phi \Rightarrow \Delta}{\Gamma \Rightarrow \Delta},$$

$$\frac{\Gamma, P : \Phi, P : \Phi \Rightarrow \Delta}{\Gamma, P : \Phi \Rightarrow \Delta}, \qquad \frac{\Gamma \Rightarrow P : \Phi, P : \Phi, \Delta}{\Gamma \Rightarrow P : \Phi, \Delta},$$

$$\frac{\Gamma, P : \Phi \Rightarrow P : \Psi, \Delta}{\Gamma, P : \Phi \backslash \Psi \Rightarrow \Delta}, \qquad \frac{\Gamma \Rightarrow P : \Phi, \Delta \qquad \Gamma, P : \Psi \Rightarrow \Delta}{\Gamma \Rightarrow P : \Phi \backslash \Psi, \Delta},$$

$$\frac{\Gamma, Q \; : \; \Phi, R \; : \; \Psi \Rightarrow \Delta \quad P \cong Q\&R}{\Gamma, P \; : \; \Phi\&\Psi \Rightarrow \Delta},$$

$$\frac{\Gamma \Rightarrow Q \; : \; \Phi, \Delta \quad \Gamma \Rightarrow R \; : \; \Psi, \Delta \quad P \cong Q\&R}{\Gamma \Rightarrow P \; : \; \Phi\&\Psi, \Delta},$$

$$\frac{\Gamma \Rightarrow Q \; : \; \Phi, R \; : \; \Psi, \Delta \quad P \cong Q+R}{\Gamma \Rightarrow P \; : \; \Phi+\Psi, \Delta},$$

$$\frac{\Gamma, Q \; : \; \Phi \Rightarrow \Delta \quad \Gamma, R \; : \; \Psi \Rightarrow \Delta \quad P \cong Q+R}{\Gamma, P \; : \; \Phi+\Psi \Rightarrow \Delta},$$

$$\frac{P \not\cong \mathrm{Nil}}{\Gamma, \, P : \mathrm{Nil} \Rightarrow \Delta}, \qquad \frac{P \cong \mathrm{Nil}}{\Gamma \Rightarrow P : \mathrm{Nil}, \Delta},$$

$$\frac{\forall \langle Q, R \rangle \in \mathrm{T(P)}.\Gamma, Q \; : \; \Phi, R \; : \; \Psi \Rightarrow \Delta}{\Gamma, P \; : \; \Phi|\Psi \Rightarrow \Delta},$$

$$\frac{\Gamma \Rightarrow Q \; : \; \Phi, \Delta \quad \Gamma \Rightarrow \Delta, R \; : \; \Psi \quad P \cong Q \mid R}{\Gamma \Rightarrow P \; : \; \Phi|\Psi, \Delta},$$

$$\frac{\forall Q \in \mathrm{T}(\alpha, \mathrm{P}).\Gamma, Q \; : \; \Phi \Rightarrow \Delta}{\Gamma, P : x*\Phi \Rightarrow \Delta}, \qquad \frac{\Gamma \Rightarrow Q \; : \; \Phi, \Delta \quad P \cong \alpha * Q}{\Gamma \Rightarrow P : x*\Phi, \Delta}.$$

where $\mathrm{T(P)}$ is a finite nonempty set $\{\langle Q, R\rangle : P \cong Q|R\}/(\cong \times \cong)$ and $\mathrm{T}(\alpha, \mathrm{P})$ is the singleton or empty set $\{Q : P \cong \alpha * Q\}/\cong$.

A provable/derivable formula is understood in the standard way.

Proposition 2.5 (soundness) *If a sequent* $\Gamma \Rightarrow \Delta$ *is derivable, then* $\mathbf{real}_C(\Gamma \Rightarrow \Delta)$.

Proof By induction of the derivation of $\Gamma \Rightarrow \Delta$. For checking we can use statements of Lemma 2.1. $\qquad\square$

Proposition 2.6 (completeness) *If* $\mathbf{real}_C(\Gamma \Rightarrow \Delta)$, *then* $\Gamma \Rightarrow \Delta$ *has a derivation*.

Proof The most cases are regarded in [25]. $\qquad\square$

About proof systems in precess algebras please see [29].

The plasmodia of *Physarum polycephalum* can simulate many forms of transfers including transport systems of different countries, such as transport networks of the USA or China [9, 12, 30, 42, 147]. The matter is that *Physarum polycephalum* implements the spatial logic [108]. Therefore it can simulate different processes—not only transporting transfers, but also business processes [130].

2.4 Physarum Illocutionary Logic

Involving *Physarum* process calculus as a programming language in analyzing intelligent processes allows us to formalize many kinds of human interactions within *Physarum* automata. For instance, it is possible to consider simpler versions of *illocutionary logic* [133] that was developed for explicating the logical nature of human speech acts and for checking the illocutionary Turing test on the medium of *Physarum polycephalum* plasmodia. This logic studies *illocutionary propositions*—verbal or non-verbal propositions which express our emotional and cognitive valuations to commit interactions. Let us show that indeed, in *Physarum* process calculus, we can logically formulate some simple human illocutionary propositions.

Suppose, Ψ is any proposition that is built up by a superposition of standard propositional logical connectives ($\wedge$, $\vee$, $\neg$, $\Rightarrow$) in the conventional way. Let V be a valuation of each propositional variable p such that $V(p) \subseteq \mathcal{L}$. We mean that $V(p)$ consists of possible worlds, where p is true. Now we can define whether a proposition Ψ is true in the event x of $\mathcal{L}$. If it is true, it is denoted by $x \models \Psi$.

$x \models p$ if x belongs to $V(p)$;
not $x \models false$, where *false* is any contradiction;
$x \models (\Psi \Rightarrow \Phi)$ iff not $x \models \Psi$ or $x \models \Phi$;
$x \models \neg\Psi$ iff $x \models (\Psi \Rightarrow false)$.

Within *Physarum* spatial logic, we can define the following three illocutionary propositions:

$\bigcirc_{eat} \Psi : := $ '*I would like to eat* Ψ'
$\bigcirc_{fear} \Psi : := $ '*I fear* Ψ'
$\bigcirc_{satisfy} \Psi : := $ '*I am satisfied by* Ψ'

These propositions have the following semantics:

$y \models \bigcirc_{eat} \Psi$ if for any process P containing $A(x)$, we have that if $x \models \Psi$, then P contains a transition $x \overset{A(x)}{\to} y$

$y \models \bigcirc_{fear} \Psi$ if for any process P containing $R(x)$, we have that if $x \models \Psi$, then P contains a transition $x \overset{R(x)}{\to} y$.

$x \models \bigcirc_{satisfy} \Psi$ if for any process P containing $C(x)$, we have that if P contains a transition $x \overset{C(x)}{\to} y$, then $y \models \Psi$.

$\diamondsuit_{eat} \Psi : := \neg(\bigcirc_{eat} (\neg\Psi))$.
$\diamondsuit_{fear} \Psi : := \neg(\bigcirc_{fear} (\neg\Psi))$.
$\diamondsuit_{satisfy} \Psi : := \neg(\bigcirc_{satisfy} (\neg\Psi))$.

As we see, we can consider the plasmodium activity as a verification of three basic illocutionary propositions: $\bigcirc_{eat}\Psi$, $\bigcirc_{fear}\Psi$, and $\bigcirc_{satisfy}\Psi$. Meanwhile, usually two or more different localizations of *Physarum polycephalum* do not compete with each other and reach the Nash equilibrium soon. For instance, their illocutionary propositions '*I would like to eat*' may be regarded as their joint proposition. The illocutionary Turing test means that we have a verification of illocutionary acts $\bigcirc_{eat}\Psi$, $\bigcirc_{fear}\Psi$, and $\bigcirc_{satisfy}\Psi$ for plasmodia of *Physarum polycephalum*.

2.5   Arithmetic Operations in Physarum Spatial Logic

We know that within process calculus we can convert expressions from λ-calculus. In particular, it means that we can consider arithmetic operations as processes. *Physarum* spatial logic is a biologized version of process calculus. Therefore, we can convert arithmetic operations into processes of *Physarum polycephalum* spatial logic.

Indeed, growing pseudopodia may represent a natural number n by the following parametric process:

$$\underline{n}(x, z):: = \underbrace{\check{x} * \check{x} * \cdots * \check{x}}_{n} * \check{z}.Nil$$

The process $\underline{n}(x, z)$ proceeds n times on an output port called the successor channel $\check{x} \in \{A_1, A_2, \ldots\} \cup \{R_1, R_2, \ldots\}$ (e.g. it is the same output of attractant) and once on the zero output port $\check{z} \in \{A_1, A_2, \ldots\} \cup \{R_1, R_2, \ldots\}$ before becoming inactive *Nil*. Recall that it is a "Church-like" encoding of numerals used first in λ-calculus.

An addition process takes two natural numbers i and j represented using the channels $x[i]$, $z[i]$ and $x[j]$, $z[j]$ and returns their sum as a natural number represented using channels $x[i + j]$, $z[i + j]$:

$Add(x[i], z[i], x[j], z[j], x[i + j], z[i + j]) : := (x[i] * \check{x}[i + j] * Add(x[i], z[i], x[j], z[j], x[i + j], z[i + j])) + z[i]*Copy(x[j], z[j], x[i + j], z[i + j])).$

A multiplication process takes two natural numbers i and j represented using the channels $x[i]$, $z[i]$ and $x[j]$, $z[j]$ and returns their multiplication as a natural number represented using channels $x[\underbrace{i + \cdots + i}_{j}]$, $z[\underbrace{i + \cdots + i}_{j}]$:

$Mult(x[i], z[i], x[j], z[j], x[i \times j], z[i \times j]) := Add(x[i], z[i], x[j], z[j], x[i + \cdots + i], z[i + \cdots + i]).$

The *Copy* process replicates the signal pattern on channels x and y on to channels u and v. It is defined as follows:

$Copy(x, y, u, v):: = (x * \check{u} * Copy(x, y, u, v) + y * \check{v} * Nil)$

As we see, within *Physarum* spatial logic, we can consider some processes as arithmetic operations. Also, we can combine several arithmetic operations within one process. Let us regard the following expression:

$$(10 + 20) \times (30 + 40)$$

An appropriate process is as follows:

Mult(*Add*($x[10]$, $z[10]$, $x[20]$, $z[20]$, $x[10 + 20]$, $z[10 + 20]$), $z[10 + 20]$, *Add* ($x[30]$, $z[30]$, $x[40]$, $z[40]$, $x[30 + 40]$, $z[30 + 40]$), $z[30 + 40]$, *Add*($x[30]$, $z[30]$, $x[70]$, $z[70]$, $x[2100]$, $z[2100]$), $z[2100]$).

Thus, in this chapter we have defined the process algebra and spatial logic on the slime mould transitions. They are theoretical frameworks for our programming of slime mould. Now, we can define deterministic machines within the *Physarum* spatial logic and then define an object-programming language for simulating the *Physarum polycephalum* behaviours.

Chapter 3
Decision Logics and Physarum Machines

3.1 Decisions on Databases and Codatabases

The slime mould can simulate many intelligent processes connecting to transporting. So, we can try to explicate a decision mechanism of *Physarum polycephalum* in building transporting networks. Let us start with some basic definitions in decision theory.

Definition 3.1 A typical model of a decision process Π identifies it using the ordered quintuple

$$(\mathbf{P, S, D, R, F}), \tag{3.1}$$

where $\mathbf{P}$ refers to the decision agent, $\mathbf{S}$ is the set of possible states of the world, $\mathbf{D}$ is the set of possible (alternative) actions (undertaken by $\mathbf{P}$) on $\mathbf{S}$, $\mathbf{R} \subset \mathbf{S}$ is the set of possible results following from the actions $\mathbf{D}$, $\mathbf{F}$ is an utility function taking values in [0, 1] and whose argument is in $\mathbf{S}$. Consequently, the subject $\mathbf{P}$ undertakes a decision (solves the decision problem) using a mapping $\mathbf{D}$ from a subset X of $\mathbf{S}$ into the set of results $\mathbf{R}$ taking into account the utility function $\mathbf{F}$ on $\mathbf{S}$, see [132].

In other words, $\mathbf{P}$ wants to achieve a result from a repertoire of possible outcomes using the action $\mathbf{D}$ based on the utility function $\mathbf{F}$. Although each element of Π deserves further elaboration, we, following current decision theory, take (3.1) as the standard decomposition of a decision process. Clearly, the actual undertaking of decisions does not conform exactly to Π. For instance, (3.1) suggests that the decision process is perfectly discrete. However, particular states of Π may be difficult to precisely separate and the entire state space appears to be continuous. On the other hand, it usually happens that decision-makers discretize their action space, for instance, to calculate, intuitively or mathematically, values of the utility function.

The set of possible states of the world $\mathbf{S}$ is represented as a database containing data and some ideas of how to process this data based on the set of available actions $\mathbf{D}$

© Springer International Publishing AG, part of Springer Nature 2019
A. Schumann and K. Pancerz, *High-Level Models of Unconventional Computations*, Studies in Systems, Decision and Control 159,
https://doi.org/10.1007/978-3-319-91773-3_3

to achieve the desired results **R**. As a consequence, the decision process Π proceeds by performing a finite comprehensible series of actions **D** on **S** in a way that the solution **R** can be reached by completing an appropriate algorithm, for instance, by implementing a sequential logic structure (IF/THEN/ELSE instructions) to apply a set of instructions in sequence from the top to the bottom of the algorithm.

Typically, decisions are divided into two groups: decisions under certainty (**P** knows what will happen in the world and **S** is well-structured) and decisions under uncertainty (**P** does not know what will happen and **S** contains uncertain items) corresponding to the reasoning of fuzzy logic. The latter category is particularly important, because most decisions in daily life and economics (perhaps the most important areas of practical human activities) involve decisions under uncertainty.

In this chapter we propose another possible model of decision making, which assumes that **P** does not know what will happen, because **S** grows rapidly. This model is suitable for the slime mould behaviour that propagates in all possible directions simultaneously. In the case of plasmodia, IF/THEN/ELSE instructions have no sense. For instance, labeled transition systems are continuously growing structures which can only be presented as databases in astatic form. So, we could only apply IF/THEN/ELSE instructions in this static form. But these data sets are expanding. Notice that continuously growing structures (trees, graphs, sets), such as the slime mould, are mathematically understood as non-well-founded sets [3] or coalgebras [72].

Let us consider an example of a growing structure that is called the game of two brokers. Two brokers at a stock exchange have appropriate expert systems which are used to support decision making. The network administrator illegally copied both expert systems and sold to each broker the expert system of his opponent. Then he tries to sell each of them the following information: "Your opponent has your expert system." Then the administrator tries to sell the information: "Your opponent knows that you have his expert system," etc. How should brokers use the information received from the administrator and what information at a given iteration is essential? So, we must make a decision based on an infinite hierarchy of decisions.

A sequential logic structure is based on IF/THEN/ELSE instructions, according to which if a condition X is true (in **S**), then we execute a set of instructions $\mathbf{D}_1$, or else another set of instructions $\mathbf{D}_2$ are processed (for example, we execute the *false* instructions when the resultant of the condition is false). Conditions involving the state of the world **S** and resultants **R** in a sequential logic structure may only have the following forms: superpositions of logical operators (AND, OR, and NOT); expressions using relational operators (such as 'greater than' or 'less than'); variables which have the values *true* or *false*; combinations of logical, relational, and mathematical operators.

Due to the sequential logic structure modelling Π, we can present sets of instructions **D** as *decision trees*, where nodes are conditions or resultants and edges are implications between conditions and resultants. Implications are treated in the Boolean way: (i) when "if A, then B" is true, then A is a subset of B; (ii) when "if A, then B" is false, then "if non-B, then non-A" is true.

A decision will thus be a choice from multiple alternatives, made with a fair degree of rationality. Let A be a finite set of possible alternatives $\mathbf{A} = \{a_1, a_2, a_3, \ldots, a_n\}$ from a database $\mathbf{S}$ and $\{g_1(\cdot), g_2(\cdot), g_3(\cdot), \ldots, g_n(\cdot)\}$ be a set of evaluation criteria for $\mathbf{F}$. Value patterns used to compare alternatives such as 'better than,' 'worse than,' 'equally good,' 'equal in value to,' 'at least as good as,' etc. are represented as binary relations which are called *preference relations*. So we can introduce the notion 'better than' ($\prec$) for each g_i, where $i = 1, \ldots, n$, to denote a *strong preference* according to g_i, the notion 'equal in value to' ($\approx$) for each g_i, to denote *indifference* according to g_i, and the notion 'at least as good as' ($\preceq$) for each g_i, to denote a *weak preference* according to g_i.

Let us notice that strong preference, indifference, and weak preference are *transitive*:

if $A \prec B$ and $B \prec C$, then $A \prec C$;
if $A \approx B$ and $B \approx C$, then $A \approx C$;
if $A \preceq B$ and $B \preceq C$, then $A \preceq C$;

Weak preference is *acyclic*:

$A \approx B$ if and only if $A \preceq B$ and $B \preceq A$.
Strong preference is also *acyclic*:
if neither $A \prec B$ nor $B \prec A$, then $A \approx B$;
$A \prec B$ if and only if $A \preceq B$ and not $B \preceq A$.

Weak preference is *reflexive*:

$A \preceq A$.

Indifference is *symmetric*:

if $A \approx B$, then $B \approx A$.

Strong preference is *asymmetric*:

if $A \prec B$, then not $B \prec A$.

In conventional decision theory it is assumed as well that each weak preference relation satisfies the formal property of *completeness*: the relation is complete if and only if for any elements A and B of $\mathbf{A}$, at least one of $A \preceq B$ or $B \preceq A$ holds. Note that the binary relations 'better than' and 'worse than' are not quite symmetrical from the psychological point of view: "A is better than B" is not exactly the same in our perceptions as "B is worse than A". For example, suppose a manager discusses the abilities of two employees. If he says "the second employee is better than the first employee," he may be satisfied with both of them, but if he says "the second employee is worse than the first employee," then he probably wants to dismiss them both from their jobs.

The preference relations are a good basis for ordering a database $\mathbf{S}$. Binary relations by which we can order entities within a database can be also understood in terms of utility relations ('gives more profit than,' 'of equal profit,' etc.), loss relations ('causes more loss than,' 'of equal loss,' etc.) and so on. A database ordered according to preference relations, utility relations, etc. may be presented as an algebraic system.

If a database can be represented as an algebraic system, then we can use conventional logics (e.g. classical logic) to make decisions: consider an ordered set (e.g. inductive set), then we can interpret logical operations as follows:

the implication $a \Rightarrow b$ is true if and only if $a \leq b$ (e.g. $a \preceq b$);
the negation $\neg a$ is true if and only if $a \Rightarrow 0$ is true, where 0 is a minimal member of the database;
the disjunction $a \vee b = c$ is true if and only if c is a minimal member such that $c \geq a$ and $c \geq b$;
the conjunction $a \wedge b = c$ is true if and only if c is a maximal member such that $a \geq c$ and $b \geq c$.

Note that $a \vee b = \neg a \Rightarrow b$ and $a \wedge b = \neg(\neg a \vee \neg b)$.

Nevertheless, in most cases we deal with uncertainty in data and cannot define sets precisely. But representing databases in the form of algebraic systems is still possible. On the one hand, in the case of uncertain entities we appeal to *bounded rationality* that captures the fact that rational choices are constrained by the limits of knowledge and cognitive capability. On the other hand, we can generalize logical (algebraic) operations to operations on uncertain entities as well.

There exist a lot of notions which are uncertain (imprecisely defined), such as 'being young', 'being tall', 'being healthy', 'being bald', etc. Fuzzy set theory and fuzzy logic reflect the fuzzy concepts and reasoning in which such items occur. A *fuzzy set* (sometimes called 'rough set') A such as 'young people' is defined by its membership function μ_A that takes values in the interval of real numbers $[0, 1]$ which indicate the degree of membership as to how imprecise elements x belong to A. If $\mu_A(x) = 1$, then it means that x certainly belongs to A. If $\mu_A(x) = 0$, then x certainly does not belong to A, and if $0 < \mu_A(x) < 1$, then x only partially belongs to A. We can define the following logical operations on a fuzzy set:

A is a subset of $B ::= \mu_A(x) \leq \mu_B(x)$;
A and $B ::= \min(\mu_A(x), \mu_B(x))$;
A or $B ::= \max(\mu_A(x), \mu_B(x))$;
non-$A ::= 1 - \mu_A(x)$.

Note that fuzzy sets differ from probabilistic sets. For instance, assume that there are two water bottles A and B. Let bottle A belong to the set of water for drinking with the membership function $= 0.9$ and bottle B belong to the set of water for drinking with probability $= 90\%$. Which bottle is preferable for drinking? A is certainly not a good choice, because it is not for drinking at 90%. B may be a good choice or not at 90%.

Hence, in everyday decisions we very often refer to fuzzy IF/THEN/ELSE reasoning such as:

If a client's profit is 'big,' then his credit rating is 'good.'

The terms 'big' and 'good' are fuzzy. For example, we can suppose that a 'big profit' means a profit of more than 5%. Let us consider another example. Assume that a trader defines a trading rule by means of a long position when the slope defining a trend is greater than or equal to a certain value x and volatility is less than or equal to a certain value y. Such a rule can be interpreted as follows: If the slope $\geq x$ and the volatility $\leq y$, then the long position should be equal to z, where x, y, z are parameters defined by the trader. The following is a fuzzy version of the same rule: If the slope is large and positive and the volatility is low, then the trading position is long.

Definition 3.2 Fuzzy databases can be modelled by the ordered set (α, T, X, G, M), where α is a linguistic variable, T is a set of its meanings (terms) representing all the names of the fuzzy variable α which are defined on the set X, G is a set of syntactic rules on T, allowing, in particular, to generate new terms (meanings of α); M is a set of semantic rules, allowing to refer each new term to meanings of the fuzzy variable α.

For example, to define the meaning of income, we can introduce the notions 'small,' 'average,' and 'big' income. Let the minimal income be equal to $2000 and maximal to $10,000. Now we can define a fuzzy database for the linguistic variable 'income' within the ordered system (α, T, X, G, M), where α is income; T = 'small income,' 'average income,' 'big income'; $X = [\$2000, \$10,000]$; G is a set of syntactic rules for generating new terms by means of the connectives 'and,' 'or,' 'not,' 'very,' etc., e.g. 'small *or* average income,' '*very* big income,' etc.; M is a set of semantic rules mapping the fuzzy subsets 'small income,' 'average income,' 'big income', as well as their logical superpositions into the set $X = [\$2000; \$10,000]$. Note that the ordered set (α, T, X, G, M) is a kind of database whose data can also be presented as an inductive set. This means that we can develop a conventional logic, which is called *fuzzy logic*, for imprecise data. Using fuzzy logic we can also make deductive decisions.

Thus, the IF/THEN/ELSE instructions for decisions in **S** can cover (i) precisely defined items, where we can deal with **S** represented as an algebraic system, or (ii) imprecisely defined items, where we can deal with **S** represented as a fuzzy database.

Now, let us construct an infinite *hierarchy* of (fuzzy) decisions in accordance with labels $l = 0, 1, 2, \ldots$ which mean that a decision with label i is more important than a decision with label j if and only if $i > j$.

Let us consider **S** as a sequence of ensembles $\mathscr{S}_l$ labelled by l (importance in the hierarchy) and having volumes $card(\mathscr{S}_l)$, $l = 0, 1, 2, \ldots$ Let $\mathscr{S} = \prod_{j=0}^{\infty} \mathscr{S}_j$. We may imagine the ensemble $\mathscr{S}$ as being the population of a tower $T = T_{\mathscr{S}}$ which has an infinite number of floors with the population of the j-th floor being $\mathscr{S}_j$. Set $T_k = \mathscr{S} = \prod_{j=0}^{\infty} \mathscr{S}_j \times \prod_{m=k+1}^{\infty} \emptyset$. This is the population of the first $k + 1$ floors.

The cardinality of T_k is defined as follows: $card(T_k) :: = (card(T_1), card(T_2), card(T_3), \ldots)$, i.e. it is a stream of cardinal numbers. The cardinality of $\mathscr{S}$ is defined thus: $card(\mathscr{S}) :: = \lim_{k \to \infty} card(T_k)$. Arithmetic operations involving the numbers $card(A)$, $card(B)$, such that $A \subseteq \mathscr{S}$ and $B \subseteq \mathscr{S}$, are calculated digit by digit: $card(A) * card(B) :: = (card(A_1) * card(B_1), card(A_2) * card(B_2), card(A_3) * card(B_3), \ldots)$, where $* \in \{+, -, \cdot, /, \inf, \sup\}$.

Let $A \subset \mathscr{S}$. We define the probability of A by the standard proportional relation:

$$P(A) :: = P_{\mathscr{S}}(A) = \frac{card(A \cap \mathscr{S})}{card(\mathscr{S})}.$$

So $P(\mathscr{S}) = \mathbf{1}$ and $P(\emptyset) = \mathbf{0}$, where $\mathbf{1} = (1, 1, 1, \ldots)$, $\mathbf{0} = (0, 0, 0, \ldots)$. If $A \subseteq \mathscr{S}$ and $B \subseteq \mathscr{S}$ are disjoint, i.e. $\inf(P(A), P(B)) = \mathbf{0}$, then $P(A \cup B) = P(A) + P(B)$. Otherwise, $P(A \cup B) = \sup(P(A), P(B))$. $P(\neg A) = \mathbf{1} - P(A)$ for all $A \subseteq \mathscr{S}$, where $\neg A = \mathscr{S} - A$.

Relative probability functions $P(A|B)$ are defined as follows:

$$P(A|B) = \frac{P(A \cap B)}{P(B)},$$

where $P(B) \neq 0$ and $P(A \cap B) = \inf(P(A), P(B))$.

Let At be a finite nonempty set of attributes which express the properties of $s \in \mathscr{S}_l$, V_a is a nonempty set of values $v \in V_a$ for $a \in At$, $I_a \colon \mathscr{S}_l \to V_a$ is an information function that maps an object in $\mathscr{S}_l$ to a value of $v \in V_a$ for an attribute $a \in At$. Now we consider all the Boolean compositions of atomic formulas $(a, v)_l$. The meaning $||\Phi_l||_{\mathscr{S}}$ of formulas Φ_l in our language is defined in the following way:

$$||(a, v)_l||_{\mathscr{S}} = \{s \in \mathscr{S}_l : I_a(s) = v\}, a \in At, v \in V_a;$$
$$||\Phi_l \vee \Psi_l||_{\mathscr{S}} = ||\Phi_l||_{\mathscr{S}} \cup ||\Psi_l||_{\mathscr{S}};$$
$$||\Phi_l \wedge \Psi_l||_{\mathscr{S}} = ||\Phi_l||_{\mathscr{S}} \cap ||\Psi_l||_{\mathscr{S}};$$
$$||\neg \Phi_l||_{\mathscr{S}} = \mathscr{S}_l - ||\Phi_l||_{\mathscr{S}}.$$

A finite hierarchy of (fuzzy) decisions is understood as follows:

$$\Phi^k :: = (\Phi_1, \Phi_2, \ldots, \Phi_k)$$

with the meaning:

$$||\Phi^k|| :: = (||\Phi_1||, ||\Phi_2||, \ldots, ||\Phi_k||)$$

An infinite hierarchy of (fuzzy) decisions is understood as follows:

$$\Phi_\infty :: = \lim_{l \to \infty} (\Phi_1, \Phi_2, \ldots, \Phi_l, \ldots)$$

with the meaning:

$$||\Phi_\infty|| :: = \lim_{l \to \infty} (||\Phi_1||, ||\Phi_2||, \ldots, ||\Phi_l||, \ldots)$$

A decision rule over decisions of different hierarchy levels is a graph $\Phi^m \to \Psi^n$, where Φ^m is a parent and Ψ^n is a child, which can be interpreted as the following appropriately defined conditional probability:

$$\pi_\mathscr{S}(\Psi^n|\Phi^m) = P(||\Psi^n||_\mathscr{S} \mid ||\Phi^m||_\mathscr{S}) = \frac{card(||\Psi^n||_\mathscr{S} \cap ||\Phi^m||_\mathscr{S})}{card(||\Phi^m||_\mathscr{S})},$$

where $||\Phi^m||_\mathscr{S} \neq \emptyset$ and $card(||\Phi^m||_\mathscr{S}) = \prod_{j=0}^{m} card(||\Phi^m||_\mathscr{S}) \times \prod_{n=m+1}^{\infty} card(\emptyset)$.

In this way we can construct Bayesian networks using Bayes' formula:

$$\pi_\mathscr{S}(\Psi^n|\Phi^m) = \frac{\pi_\mathscr{S}(\Psi^n) \cdot \pi_\mathscr{S}(\Phi^m|\Psi^n)}{\pi_\mathscr{S}(\Phi^m|\Psi^n) \cdot \pi_\mathscr{S}(\Psi^n) + \pi_\mathscr{S}(\Phi^m|\neg\Psi^n) \cdot \pi_\mathscr{S}(\neg\Psi^m)},$$

where $\pi_\mathscr{S}(\Psi^n|\Phi^m)$ is the *a posteriori* probability of Ψ^n given Φ^m, $\pi_\mathscr{S}(\Psi^n)$ is the *a priori* probability of Ψ^n, and $\pi_\mathscr{S}(\Phi^m|\Psi^n)$ is the likelihood of Φ^m given Ψ^n.

Hence, we can build Bayesian networks on **S**, represented as a hierarchy of ensembles, to make a decision over an (infinite) hierarchy of different decisions.

Precise or fuzzy data have to be fixed, i.e. they are limited by inductions (least fixed points), so that the number of their members does not change. Such data satisfy the set-theoretic axiom of foundation (e.g. this means that they are inductive sets) and hence such data are called *well-founded*. Nevertheless, in the case of decisions over a hierarchy of other decisions,we have dealt with hierarchies that change continuously, e.g. grow rapidly. Such an infinite hierarchy does not satisfy the foundation axiom and therefore involve so called *non-well-founded data* or *codata* [3]. Making decisions involving codata is much more sophisticated than making decisions involving well-founded data, because we cannot formulate algorithmic decision rules.

The most natural examples of codata result from the slime mould behaviour. A process in the plasmodium transition is a state-based system transforming an input sequence into an output sequence. It obtains step-by-step an input value and, depending on its current state, it produces an output value and changes its state. According to this definition, no process can be fixed as an inductive set. It is always changing, i.e. it flows, as Heraclitus of Ephesus would say. Mathematically, a process is given as a coalgebra by the following entities [72]: a set S of states, a set I of inputs, a set O of outputs, a set R of results, and a function $f : S \times I \to C(R + S \times O)$, for some functor C. The function f describes one step of the process: in the state $s \in S$ and the input $i \in I, f$ chooses a possible continuation that consists of either terminating with a result $r \in R$, or continuing in a state $s' \in S$ and producing an output value $o \in O$.

A *coalgebra* can be regarded as a greatest fixed point of the choice functor C. This functor C determines which kind of process we have. Important examples for choice functors are as follows:

- the deterministic choice functor C^{det} represented by the *identity functor*, which is used if the input uniquely determines what happens next;
- a non-deterministic choice functor C^{ndet} represented by a *finite power-set functor*, which is used if, for a given input, there may be various possible continuations of the process;
- a probabilistic choice functor C^{prob} represented by a *finite probability functor*, which is used if the continuation of the process is random.

Thus, codata are a process defined coalgebraically, where the choice functor can be understood in various ways. We transform the data $\mathbf{S}$ into an appropriate *codatabase* when we know that an unconventional logic is suited to making deductive decisions on codata of $\mathbf{S}$. So it is possible to deal not only with (fuzzy) databases, but also with codatabases.

Graphically, coalgebras (e.g. processes or games) can be represented as infinite trees. Trees defined coalgebraically are called *non-well-founded*. They are the simplest graphic examples of codata.

Now let us try to define fuzzy reasoning on codatabases. Each codatabase can be considered as a transition system $TS = (S, E, T, I)$, where:

1. S is a non-empty set of states,
2. E is the set of actions,
3. $T \subseteq S \times E \times S$ is the transition relation,
4. $I \subseteq S$ is the set of initial states.

In transition systems, transitions are performed by labeled actions in the following way: if $(s, e, s') \in T$, then the system goes from s to s'. The element $(s, e, s') \in T$ is called a transition.

Any transition system $TS = (S, E, T, I)$ can be represented in the form of a labeled graph with nodes corresponding to states of S, edges representing the transition relation T, and labels of edges corresponding to events of E.

3.2   Determenistic Transition Systems

Usually, machines are understood determenistic in applying a transition rule: the values of the next states are being obtained by functions defined on the values of the current states. In case of transition systems, the terms "deterministic" and "non-deterministic" have been used differently in some of the literature. Typically, the term "deterministic" refers to transitions systems which are "monogenic". By monogenic, we mean a transition system $TS = (S, E, T, S_{init})$ such that for each state $s \in S$, there is at most one $s' \in S$ such that $s \xrightarrow{e} s'$ for any $e \in E$ (cf. [44]). In this book, we adopt this definition of deterministic transition systems.

Let $TS = (S, E, T, S_{init})$ be a transition system. For each state $s \in S$ of TS, we can determine its direct successors and predecessors. Let:

- $Post(s, e) = \{s' \in S : (s, e, s') \in T\}$,
- $Pre(s, e) = \{s' \in S : (s', e, s) \in T\}$,

then the set $Post(s)$ of all direct successors of the state $s \in S$ is given by

$$Post(s) = \bigcup_{e \in E} Post(s, e)$$

and the set $Pre(s)$ of all direct predecessors of the state $s \in S$ is given by

$$Pre(s) = \bigcup_{e \in E} Pre(s, e).$$

The set $Post(s)$ is also called the postcondition of p, whereas the set $Pre(s)$ is also called the precondition of p.

For a given state $s \in S$, all states included in $Post(s)$ are called the states directly reachable from the state s. If $Post(s) = \emptyset$, then s is said to be a goal state in the transition system TS.

Example 3.1 Let us consider a transition system $TS = (S, E, T, S_{init})$ given in Example 2.1. We obtain the following direct successors and predecessors for states in TS:

- $Post(s_1) = \{s_2, s_3, s_4\}, Pre(s_1) = \emptyset$,
- $Post(s_2) = \{s_5\}, Pre(s_2) = \{s_1\}$,
- $Post(s_3) = \emptyset, Pre(s_3) = \{s_1\}$,
- $Post(s_4) = \emptyset, Pre(s_4) = \{s_1\}$,
- $Post(s_5) = \emptyset, Pre(s_5) = \{s_2\}$.

The states s_3, s_4, and s_5 are the goal states.

3.3 Timed Transition System

It is assumed, in transition systems mentioned earlier, that all events happen instantaneously. In timed transition systems, timing constraints restrict the times at which events may occur [38]. The timing constraints are classified into two categories: lower-bound and upper-bound requirements.

Definition 3.3 Let N be a set of nonnegative integers. A timed transition system $TTS = (S, E, T, S_{init}, l, u)$ consists of:

- an underlying transition system $TS = (S, E, T, S_{init})$,
- a minimal delay function (a lower bound) $l : E \rightarrow N$ assigning a nonnegative integer to each event,
- a maximal delay function (an upper bound) $u : E \rightarrow N \cup \{\infty\}$ assigning a non-negative integer or infinity to each event.

Remark 3.1 In this book, we assume, for timed transition systems, that the events may occur only at discrete time instants. Therefore, whenever time instant t is used, it means that $t = t_0, t_1, t_2, \ldots$. See [38, 118, 123].

Example 3.2 Let us add some timing constraints to events in a transition system $TS = (S, E, T, S_{init})$ given in Example 2.1. We obtain a timed transition system $TTS = (S, E, T, S_{init}, l, u)$, presented in the form of a labeled directed graph in Fig. 3.1, where:

- $S = \{s_1, s_2, s_3, s_4, s_5\}$,
- $E = \{e_1, e_2, e_3, e_4\}$,
- $T = \{(s_1, e_1, s_2), (s_1, e_2, s_3), (s_1, e_3, s_4), (s_2, e_4, s_5)\}$,
- $S_{init} = \{s_1\}$.
- $l(e_1) = l(e_3) = l(e_4) = 0, l(e_2) = 5$,
- $u(e_1) = u(e_3) = u(e_4) = \infty, u(e_2) = 10$.

Let $TTS = (S, E, T, S_{init}, l, u)$ be a timed transition system. For each state $s \in S$ in TTS and each $t \in \{t_0, t_1, t_2, \ldots\}$, we can determine its direct successors and predecessors at the time instant t. Let

$$Post_t(s, e) = \begin{cases} \{s' \in S : (s, e, s') \in T\} & \text{if } l(e) \leq t \leq u(e), \\ \emptyset & \text{otherwise} \end{cases}$$

and

$$Pre_t(s, e) = \begin{cases} \{s' \in S : (s', e, s) \in T\} & \text{if } l(e) \leq t \leq u(e), \\ \emptyset & \text{otherwise} \end{cases}$$

Fig. 3.1 The timed transition system TTS presented in the form of a labeled directed graph

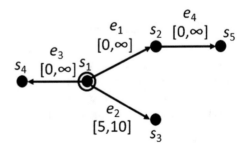

then the set $Post_t(s)$ of all direct successors of the state $s \in S$ at t is given by

$$Post_t(s) = \bigcup_{e \in E} Post_t(s, e)$$

and the set $Pre_t(s)$ of all direct predecessors of the state $s \in S$ at t is given by

$$Pre_t(s) = \bigcup_{e \in E} Pre_t(s, e).$$

Example 3.3 Let us consider a timed transition system $TTS = (S, E, T, S_{init}, l, u)$ given in Example 3.3. We obtain the following direct successors and predecessors for states in TTS:

- $Post_t(s_1) = \{s_2, s_4\}$ if $t < 5$ or $t > 10$, $Post_t(s_1) = \{s_2, s_3, s_4\}$ if $t \geq 5$ and $t \leq 10$, $Pre_t(s_1) = \emptyset$ for each t,
- $Post_t(s_2) = \{s_5\}$ and $Pre_t(s_2) = \{s_1\}$ for each t,
- $Post_t(s_3) = \emptyset$ for each t, $Pre_t(s_3) = \emptyset$ if if $t < 5$ or $t > 10$, $Pre_t(s_3) = \{s_1\}$ if $t \geq 5$ and $t \leq 10$,
- $Post_t(s_4) = \emptyset$ and $Pre_t(s_4) = \{s_1\}$ for each t,
- $Post_t(s_5) = \emptyset$ and $Pre_t(s_5) = \{s_2\}$ for each t.

3.4 Physarum Machines

A *Physarum* machine is a biological computing device experimentally implemented in the plasmodium of *Physarum polycephalum*. The *Physarum* machine comprises an amorphous yellowish mass (see Fig. 3.2) with networks of protoplasmic veins, programmed by spatial configurations of attracting and/or repelling stimuli. When attractants are scattered in the plasmodium range, a network of protoplasmic veins, connecting the original points of plasmodium and those attractants, is formed. The plasmodium looks for attractants, propagates protoplasmic veins towards them, feeds on them and goes on. As a result, a natural transition system is built up (see [4, 108]). Each original point of plasmodium and each attractant occupied by plasmodium is called an active point in the *Physarum* machines. Activated repellents can avoid or anihilate propagation of protoplasmic veins towards activated attractants.

Formally, a structure of the *Physarum* machine can be described as a triple $\mathscr{PM} = (Ph, Attr, Rep)$ (cf. [61]), where:

- $Ph = \{ph_1, ph_2, \ldots, ph_k\}$ is the set of original points of plasmodium,
- $Attr = \{attr_1, attr_2, \ldots, attr_m\}$ is the set of attractants;
- $Rep = \{rep_1, rep_2, \ldots, rep_n\}$ is the set of repellents.

In a standard case, positions of original points of plasmodium, attractants, and repellents are considered in the two-dimensional space (for example, at a Petri dish [68]).

Example 3.4 Let us consider a structure $\mathscr{PM} = (Ph, Attr, Rep)$ of the *Physarum* machine given in Fig. 3.3. It is worth noting that, in the graphical presentation of structures of *Physarum* machines, we will use the following symbols:

- filled circles corresponding to original points of plasmodium,
- empty circles corresponding to attractants,
- empty rectangles corresponding to repellents.

One can see that the components of the structure $\mathscr{PM} = (Ph, Attr, Rep)$ are as follows:

- $Ph = \{ph\}$,
- $Attr = \{attr_1, attr_2, attr_3, attr_4, attr_5, attr_6, attr_7\}$,
- $Rep = \{rep\}$.

In general, a dynamics (behaviour) of the *Physarum* machine $\mathscr{PM}$ can be described by the family $V = \{V^t\}_{t \in \{t_0, t_1, t_2, \ldots\}}$ of the sets of protoplasmic veins formed by plasmodium during its action, where $V^t = \{v_1^t, v_2^t, \ldots, v_{card(V^t)}^t\}$ is the set of all protoplasmic veins of plasmodium present at the time instant t in $\mathscr{PM}$. Each vein $v_i^t \in V^t$, where $i = 1, 2, \ldots, card(V^t)$, is the pair $\langle \pi_{i_s}^t, \pi_{i_e}^t \rangle$ of active points in $\mathscr{PM}$, i.e., $\pi_{i_s}^t \in Ph \cup Attr$ and $\pi_{i_e}^t \in Ph \cup Attr$. $\pi_{i_s}^t$ is the start point of the vein v_i^t whereas $\pi_{i_e}^t$ is the end point of the vein v_i^t.

The starting point in modeling behaviour of a given *Physarum* machine $TS(\mathscr{PM})$ is a transition system describing plasmodium propagation (see [61]). To build a model, in the form of a transition system $TS(\mathscr{PM}) = (S, E, T, S_{init})$, of behaviour of the *Physarum* machine $\mathscr{PM} = \{Ph, Attr, Rep\}$, we take into consideration a stable state, i.e., the state at a given time instant t (for example, the last one), when the set of all protoplasmic veins formed by plasmodium is fixed, i.e., $V = \{v_1, v_2, \ldots, v_{card(V)}\}$ (note that the superscript t has been omitted). The following bijective functions are used:

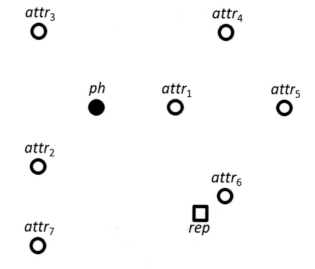

Fig. 3.3 A structure
$\mathscr{PM} = (Ph, Attr, Rep)$ of
the *Physarum* machine

- $\sigma : Ph \cup Attr \rightarrow S$ assigning a state to each original point of plasmodium as well as to each attractant,
- $\varepsilon : V \rightarrow E$ assigning an event to each protoplasmic vein,
- $\tau : V \rightarrow T$ assigning a transition to each protoplasmic vein,
- $\iota : Ph \rightarrow S_{init}$ assigning an initial state to each original point of plasmodium.

Example 3.5 Let us consider a stable state of the *Physarum* machine $\mathscr{PM} = (Ph, Attr, Rep)$ from Example 3.4 as in Fig. 3.4. One can see that protoplasmic veins were formed by plasmodium. A model, in the form of a transition system $TS(\mathscr{PM}) = (S, E, T, S_{init})$, presents a behaviour of the *Physarum* machine $\mathscr{PM} = \{Ph, Attr, Rep\}$, where:

- $S = \{s_1, s_2, s_3, s_4, s_5, s_6, s_7, s_8\}$,
- $E = \{e_1, e_2, e_3, e_4, e_5, e_6, s_7, s_8\}$,
- $T = \{(s_1, e_1, s_2), (s_1, e_2, s_3), (s_1, e_3, s_4), (s_2, e_4, s_5), (s_2, e_5, s_6), (s_2, e_6, s_7), (s_3, e_7, s_8)\}$,
- $S_{init} = \{s_1\}$,

is shown in Fig. 3.5. The bijective functions for building a model in the form of $TS(\mathscr{PM})$ are as follows:

- $\sigma(ph) = s_1, \sigma(attr_1) = s_2, \sigma(attr_3) = s_3, \sigma(attr_3) = s_4, \sigma(attr_4) = s_5, \sigma(attr_5) = s_6, \sigma(attr_6) = s_7, \sigma(attr_7) = s_8$,
- $\varepsilon(\langle ph, attr_1\rangle) = e_1, \varepsilon(\langle ph, attr_2\rangle) = e_2, \varepsilon(\langle ph, attr_3\rangle) = e_3, \varepsilon(\langle attr_1, attr_4\rangle) = e_4, \varepsilon(\langle attr_1, attr_5\rangle) = e_5, \varepsilon(\langle attr_1, attr_6\rangle) = e_6, \varepsilon(\langle attr_2, attr_7\rangle) = e_7$,

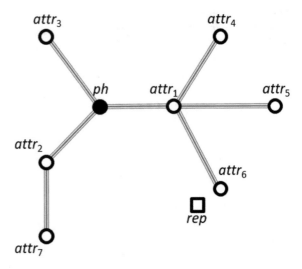

Fig. 3.4 A stable state of the *Physarum* machine $\mathscr{P}\mathscr{M} = (Ph, Attr, Rep)$

- $\tau(\langle ph, attr_1\rangle) = (s_1, e_1, s_2)$, $\tau(\langle ph, attr_2\rangle) = (s_1, e_2, s_3)$, $\tau(\langle ph, attr_3\rangle) = (s_1, e_3, s_4)$, $\tau(\langle attr_1, attr_4\rangle) = (s_2, e_4, s_5)$, $\tau(\langle attr_1, attr_5\rangle) = (s_2, e_5, s_6)$, $\tau(\langle attr_1, attr_6\rangle) = (s_2, e_6, s_7)$, $\tau(\langle attr_2, attr_7\rangle) = (s_3, e_7, s_8)$,
- $\iota(ph) = s_1$.

We can identify in the *Physarum* machine $\mathscr{P}\mathscr{M}$ five full paths of plasmodium propagation. These paths are determined by strings of events in the transition system model $TS(\mathscr{P}\mathscr{M})$ of $\mathscr{P}\mathscr{M}$, i.e.:

- $s_1 \xrightarrow{e_1} s_2 \xrightarrow{e_4} s_5$,
- $s_1 \xrightarrow{e_1} s_2 \xrightarrow{e_5} s_6$,
- $s_1 \xrightarrow{e_1} s_2 \xrightarrow{e_6} s_7$,
- $s_1 \xrightarrow{e_2} s_3 \xrightarrow{e_7} s_8$,
- $s_1 \xrightarrow{e_3} s_4$.

In [123], timed transition systems were used to model the behaviour of *Physarum* machines. In timed transition systems, the quantitative lower-bound and upper-bound timing constraints are imposed on events. This ability of modeling the behaviour of *Physarum* machines is important because attracting and repelling stimuli can be activated and/or deactivated for proper time periods to perform given computational tasks. In case of a model in the form of a timed transition system $TTS(\mathscr{P}\mathscr{M}) = (S, E, T, S_{init}, l, u)$, the bijective functions are slightly modified, i.e.:

- $\sigma : Ph \cup Attr \to S$ assigning a state to each original point of plasmodium as well as to each attractant,
- $\epsilon : \left(\bigcup_{t\in\{t_0,t_1,t_2,\dots\}} V_t \right) \to E$ assigning an event to each protoplasmic vein,

Fig. 3.5 A model, in the
form of a transition system
$TS(\mathscr{PM}) = (S, E, T, S_{init})$,
of behaviour of the
Physarum machine
$\mathscr{PM} = \{Ph, Attr, Rep\}$

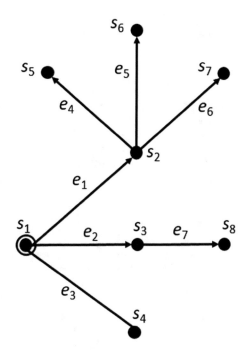

- $\tau : \left(\bigcup\limits_{t \in \{t_0, t_1, t_2, \ldots\}} V_t \right) \rightarrow T$ assigning a transition to each protoplasmic vein,
- $\iota : Ph \rightarrow S_{init}$ assigning an initial state to each original point of plasmodium.

Delay functions l and u are determined on the basis of information on time instants when protoplasmic veins were formed.

Example 3.6 Let us consider the *Physarum* machine $\mathscr{PM} = (Ph, Attr, Rep)$ from Example 3.4 at four time instants $t_0 = 0$, $t_1 = 1$, $t_2 = 2$, and $t_3 = 3$ as in Fig. 3.6. One can see that one of protoplasmic veins (between attractants $attr_1$ and $attr_6$) were anihilated because of activation of the repellent rep. Now, a model (shown in Fig. 3.5) of behaviour of the *Physarum* machine $\mathscr{PM} = \{Ph, Attr, Rep\}$ has the form of a timed transition system $TTS(\mathscr{PM}) = (S, E, T, S_{init}, l, u)$, where:

- $S = \{s_1, s_2, s_3, s_4, s_5, s_6, s_7, s_8\}$,
- $E = \{e_1, e_2, e_3, e_4, e_5, e_6, s_7, s_8\}$,
- $T = \{(s_1, e_1, s_2), (s_1, e_2, s_3), (s_1, e_3, s_4), (s_2, e_4, s_5), (s_2, e_5, s_6), (s_2, e_6, s_7),$
  $(s_3, e_7, s_8)\}$,
- $S_{init} = \{s_1\}$,
- $l(e_1) = l(e_2) = l(e_3) = l(e_4) = l(e_5) = l(e_7) = 0, l(e_6) = 1$,
- $u(e_1) = u(e_2) = u(e_3) = u(e_4) = u(e_5) = u(e_7) = \infty, u(e_6) = 2$.

We leave the reader with determining the bijective functions for building a model in the form of $TTS(\mathscr{PM})$ Fig. 3.7.

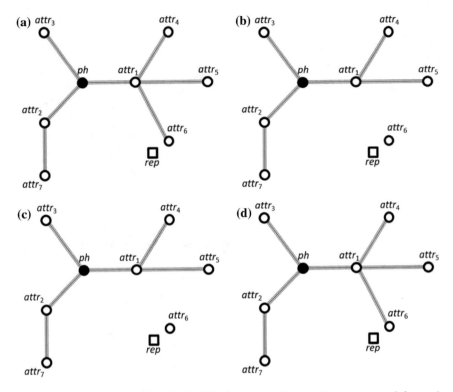

Fig. 3.6 The *Physarum* machine $\mathscr{PM} = (Ph, Attr, Rep)$ at four time instants: **a** $t_0 = 0$, **b** $t_1 = 1$, **c** $t_2 = 2$, **d** $t_3 = 3$

We can define the following four basic forms of motions in *Physarum* machines:

- *direct*—direction, i.e. a movement of plasmodium from one point, where it is located, towards another point, where there is a neighbouring attractant,
- *fuse*—fusion of two protoplasmic veins of plasmodium at the point, where they meet the same attractant,
- *split*—splitting plasmodium from one active point into two active points, where two neighbouring attractants with a similar power of intensity are located,
- *repel*—repelling of plasmodium or inaction.

Each of four basic forms of motions in *Physarum* machines can be modeled by transition systems. In Fig. 3.8a–d, transition system models of four basic forms of motions in *Physarum* machines, are shown. Formally, we have:

- *direct*: $TS^d(\mathscr{PM}) = (S^d, E^d, T^d, S_{init}^d)$, where:

  - $S^d = \{s_1, s_2\}$,
  - $E^d = \{e_1\}$,

Fig. 3.7 A model, in the form of a timed transition system 3.5, of behaviour of the *Physarum* machine $\mathscr{PM} = \{Ph, Attr, Rep\}$

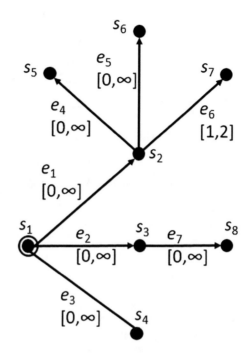

$$- T^d = \{(s_1, e_1, s_2)\},$$
$$- S^d_{init} = \{s_1\},$$

see Fig. 3.8a,

- *fuse*: $TS^f(\mathscr{PM}) = (S^f, E^f, T^f, S^f_{init})$, where:

$$- S^f = \{s_1, s_2, s_3\},$$
$$- E^f = \{e_1, e_2\},$$
$$- T^f = \{(s_1, e_1, s_3), (s_2, e_2, s_3)\},$$
$$- S^f_{init} = \emptyset,$$

see Fig. 3.8b,

- *split*: $TS^s(\mathscr{PM}) = (S^s, E^s, T^s, S^s_{init})$, where:

$$- S^s = \{s_1, s_2, s_3\},$$
$$- E^s = \{e_1, e_2\},$$
$$- T^s = \{(s_1, e_1, s_2), (s_1, e_2, s_3)\},$$
$$- S^s_{init} = \emptyset,$$

see Fig. 3.8c,

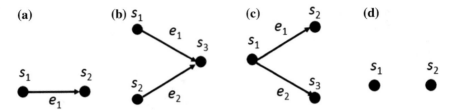

Fig. 3.8 Models in the form of transition systems for: **a** *direct*, **b** *fuse*, **c** *split*, and **d** *repel*

- *repel*: $TS^f(\mathscr{PM}) = (S^f, E^f, T^f, S_{init}^f)$, where:
  - $S^r = \{s_1, s_2\}$,
  - $E^r = \emptyset$,
  - $T^r = \emptyset$,
  - $S_{init}^r = \emptyset$,

  see Fig. 3.8d.

One can see that we have set $S_{init} = \emptyset$ for each model. It is not obligatory. We have assumed that basic forms of motions can appear in any parts of *Physarum* machines, not only in their initial parts (i.e. parts where plasmodium starts propagating). Naturally, *fuse* and *split* motions can be considered in general case when a number of protoplasmic veins of plasmodium meet the same attractant or plasmodium is split from one active point into a number of active points, respectively.

Example 3.7 Let us consider the *Physarum* machine $\mathscr{PM}$ from Example 3.5. It is easy to see, for instance, the *split* motion. Plasmodium is split from one active point (the attractant $attr_1$) into three active points (the attractants: $attr_4$, $attr_5$, and $attr_6$).

3.5 Decision Logics of Transition Systems

Let us extend the transition system *TS* to the following information system: $\mathscr{S} = (S, At, \{V_a : a \in At\}, \{I_a : a \in At\})$, where

- S is a finite nonempty set of objects called universe associated with the set of all states of *TS*;
- At is a finite nonempty set of attributes which express properties of $s \in S$ (such as color, intensity, chemical formula of attractants);
- V_a is a nonempty set of values $v \in V_a$ for $a \in At$ (such as type of color, type of intensity, type of chemical formula, etc.);
- $I_a : S \to V_a$ is an information function that maps an object in S to a value of $v \in V_a$ for an attribute $a \in At$ (e.g. this attractant $s \in S$ is blue).

Now, let us build a standard logical language $\mathscr{L}_S$ closed over Boolean compositions of atomic formulas (a, v). The meaning $||\Phi||_S$ of formulas $\Phi \in \mathscr{L}_S$ is defined by induction:

$$||(a, v)||_S = \{s \in S : I_a(s) = v\}, a \in At, v \in V_a;$$

$$||\Phi \vee \Psi||_S = ||\Phi||_S \cup ||\Psi||_S;$$

$$||\Phi \wedge \Psi||_S = ||\Phi||_S \cap ||\Psi||_S;$$

$$||\neg\Phi||_S = S - ||\Phi||_S.$$

In the language $\mathscr{L}_S$, we can define decision rules in S as follows. Assume, each formula $\Phi \in \mathscr{L}_S$ is considered a node of directed, acyclic graph. Then a decision rule in S is a graph $\Phi \rightarrow \Psi$, where Φ is a parent and Ψ is a child, that is interpreted as an appropriate conditional probability (cf. [65]):

$$\pi_S(\Psi|\Phi) = p_S(||\Psi||_S \mid ||\Phi||_S) = \frac{card(||\Psi||_S \cap ||\Phi||_S)}{card(||\Phi||_S)},$$

where $||\Phi||_S \neq \emptyset$.

In this way, the direct cause $\{\Phi \rightarrow \Psi\}$ in decision is expressed by $\pi_S(\Psi|\Phi)$, the indirect cause $\{\Phi \rightarrow \Psi, \Psi \rightarrow \Theta\}$ by $\pi_S(\Psi|\Phi) \cdot \pi_S(\Theta|\Psi)$, the common cause $\{\Phi \rightarrow \Psi, \Theta \rightarrow \Psi\}$ by $\pi_S(\Psi|\Phi, \Theta)$, the common effect $\{\Phi \rightarrow \Psi, \Phi \rightarrow \Theta\}$ by $\pi_S(\Psi|\Phi) \cdot \pi_S(\Theta|\Phi)$.

For each formula $\Phi \in \mathscr{L}_S$ with k atomic parents, we have 2^k rows for the combinations of parent values $v \in V_a$. Each row gives a number $p \in [0, 1]$ if Φ is true, and it gives a number $1 - p$ if Φ is false. If each formula has no more than k parents, the complete network requires $O(n \cdot 2^k)$ numbers.

So, we can construct Bayesian networks in $\mathscr{L}_S$ by using the following Bayes' formula:

$$\pi_S(\Psi|\Phi) = \frac{\pi_S(\Psi) \cdot \pi_S(\Phi|\Psi)}{\pi_S(\Phi|\Psi) \cdot \pi_S(\Psi) + \pi_S(\Phi|\neg\Psi) \cdot \pi_S(\neg\Psi)},$$

where $\pi_S(\Psi|\Phi)$ is the *a posteriori* probability of Ψ given Φ, $\pi_S(\Psi)$ is the a priori probability of Ψ, and $\pi_S(\Phi|\Psi)$ is the likelihood of Φ with respect to Ψ. Hence, the Bayes' formula allows us to infer the *a posteriori* probability $\pi_S(\Psi|\Phi)$ from the *a priori* probability $\pi_S(\Psi)$ through the likelihood $\pi_S(\Phi|\Psi)$.

Our main assumption in building Bayesian networks is as follows: for all formulas $\Phi_1, \Phi_2, \ldots, \Phi_n$ involved in constructing the decision graph, there is the full joint distribution defined as the product of the local conditional distributions: $\pi_S(\Phi_1, \ldots, \Phi_n) = \prod_{i=1}^{n} \pi_S(\Phi_i|parents(\Phi_i))$, where $parents(\Phi_i)$ are all parents for Φ_i.

In defining Bayesian networks, we appeal to the following probabilities:

$$p_S(s \in S) = 1/card(S);$$

$$p_S(A \subseteq S) = \frac{card(A)}{card(S)};$$

$$\pi_S(\Psi) = p_S(||\Psi||_S).$$

Nevertheless, we can use probabilities defined on rough sets, as well. The lower predecessor anticipation $Pre_*(X)$ of X is defined as

$$Pre_*(X) = \{s \in S : Post(s) \neq \emptyset \text{ and } Post(s) \subseteq X\}.$$

The upper predecessor anticipation $Pre^*(X)$ of X is defined as

$$Pre^*(X) = \{s \in S : Post(s) \cap X \neq \emptyset\}.$$

$$p_{S_*}(A \subseteq S) = \frac{card(Pre_*(A))}{card(S)};$$

$$p_S{}^*(A \subseteq S) = \frac{card(Pre^*(A))}{card(S)}.$$

Thus, if the anticipation of states from A is exact, then $p_{S_*}(A) = p_S{}^*(A) = p_S(A)$. Some other useful properties of probabilities on rough sets given $A, B \subseteq S$ are as follows:

$$p_{S_*}(A) \leq p_S(A) \leq p_S{}^*(A);$$

$$p_{S_*}(\emptyset) = p_S{}^*(\emptyset) = 0;$$

$$p_{S_*}(S) = p_S{}^*(S) = 1;$$

$$p_{S_*}(S - A) = 1 - p_S{}^*(A);$$

$$p_S{}^*(S - A) = 1 - p_{S_*}(A);$$

$$p_{S_*}(A \cup B) \geq p_{S_*}(A) + p_{S_*}(B) - p_{S_*}(A \cap B);$$

$$p_S{}^*(A \cup B) \leq p_S{}^*(A) + p_S{}^*(B) - p_S{}^*(A \cap B).$$

We can define Bayesian networks on these probabilities, too, in the way proposed by [149]. Instead of two-valued decisions whereas Φ is true or false, they have offered to use the three regions corresponding to a three-way decision of acceptance, deferment, and rejection. For these networks, we use rough probabilities:

$$P_S(A) = [p_{S_*}(A), p_S{}^*(A)].$$

Obviously, $p_{S_*}(A) = p_S{}^*(A)$ implies $P_S(A) = p_S(A)$. If $P_S(||\Phi||_S) = 1$, then we deal with acceptance, if $P_S(||\Phi||_S) = 0$, we deal with rejection, and if $0 < P_S(||\Phi||_S) < 1$, we deal with deferment.

Hence, the slime mould behaviour is a kind of machine that is called by us the *Physarum* machine. This machine can be defined as a monogenic timed transition system. It is a rapidly growing structure that can be examined just coalgebraically as a codatabase. Nevertheless, there are possible Bayesian decisions (such as Bayesian networks) on this codatabase, too. It means that the slime mould makes rational decisions in the meaning of decision theory, indeed.

Chapter 4
Petri Net Models of Plasmodium Propagation

4.1 The Rudiments of Petri Nets

The slime mould behaviour can be examined as rational from the standpoint of decision theory. So, we can try to implement some strong mathematical tools on the slime mould transitions, such as Petri nets, cf. [119, 121]. Let us remember that *Petri nets* were developed by C. A. Petri [67] as a graphical and mathematical tool, among others, for describing information processing systems. As a graphical tool, Petri nets can be used as a visual-communication aid. As a mathematical tool, Petri nets are represented, among others, by algebraic equations. For detailed information on Petri nets, we refer the readers to [69].

We can involve two steps in the process of defining a Petri net. First, we define a Petri net structure. Then, we add an initial marking function assigning tokens to places of the Petri net.

A Petri net structure has two types of nodes, places and transitions, and arcs connecting them. It is a bipartite graph, i.e. arcs cannot directly connect nodes of the same type.

Definition 4.1 A Petri net structure is a weighted directed bipartite graph $PNS = (Pl, Tr, Arc, w)$, where:

- Pl is the finite set of places,
- Tr is the finite set of transitions,
- $Arc \subseteq (Pl \times Tr) \cup (Tr \times Pl)$ is the set of arcs from places to transitions and from transitions to places,
- $w : Arc \rightarrow \{1, 2, 3, \ldots\}$ is the weight function on the arcs.

We assume that a Petri net structure PNS has no isolated places and transitions. By $w(n_i, n_j)$ we denote the weight of the arc $(n_i, n_j) \in Arc$, where n_i is the i-th node (transition or place) in PNS and n_j is the j-th node (place or transition) in PNS.

© Springer International Publishing AG, part of Springer Nature 2019
A. Schumann and K. Pancerz, *High-Level Models of Unconventional Computations*, Studies in Systems, Decision and Control 159,
https://doi.org/10.1007/978-3-319-91773-3_4

Two types of nodes in the Petri net structure are differentiated when it is drawn. The convention is to use circles to represent places and rectangles to represent transitions.

Let $PNS = (Pl, Tr, Arc, w)$ be a Petri net structure. It is convenient to use for any $t \in Tr$:

- $I(t) = \{p \in Pl : (p, t) \in Arc\}$ is a set of input places connected through arcs to the transition t,
- $O(t) = \{p \in Pl : (t, p) \in Arc\}$ is a set of output places connected through arcs from the transition t.

Definition 4.2 A marked Petri net is a five-tuple $MPN = (Pl, Tr, Arc, w, m_0)$ consisting of:

- the Petri net structure $PNS = (Pl, Tr, Arc, w)$,
- the initial marking function $m_0 : Pl \rightarrow \{0, 1, 2, \ldots\}$ on the places.

An initial marking function m_0 assignes tokens to places. The number of tokens assigned to a given place $p \in Pl$ by m_0 is an arbitrary non-negative integer, not necessarily bounded. In the Petri net graph, a token is indicated by a dark dot positioned in the appriopriate place.

The Petri net dynamics is given by firing enabled transitions causing the movement of tokens through the net. A mapping $m : Pl \rightarrow \{0, 1, 2, \ldots\}$ assigning tokens to places is called a marking of the marked Petri net $MPN = (Pl, Tr, Arc, w, m_0)$.

Definition 4.3 Let $MPN = (Pl, Tr, Arc, w, m_0)$ be a marked Petri net. A transition $t \in Tr$ is said to be enabled if and only if $m(p) \geq w(p, t)$ for all $p \in I(t)$.

When a transition $t \in Tr$ is enabled, we say that it can fire. After firing an enabled transition t, we obtain a new marking of MPN. Let $MPN = (Pl, Tr, Arc, w, m_0)$, where $Pl = \{p_1, p_2, \ldots, p_k\}$ and $Tr = \{t_1, t_2, \ldots, t_l\}$, be a marked Petri net. A marking of MPN can be viewed as a marking row vector

$$\mathbf{m} = [m(p_0), m(p_1), \ldots, m(p_k)].$$

We can specify the next marking $\mathbf{m}'$ of MPN when a given transition $t \in Tr$ fires. To do so, we define:

- A firing row vector
$$\mathbf{f} = [f(t_1), f(t_2), \ldots, f(t_l)],$$

where 1 appears only in the i-th position, $i = 1, 2, \ldots, l$, i.e. $f(t_i) = 1$, to indicate the fact that i-th transition is currently firing. For the remainig positions, i.e. for any $j = 1, 2, \ldots, i - 1, i + 1, \ldots, l$, we assign $f(t_j) = 0$.

- An incidence matrix

$$\mathbf{A} = \begin{bmatrix} a_{00} \ a_{01} \ \cdots \ a_{0k} \\ a_{10} \ a_{11} \ \cdots \ a_{1k} \\ \cdots \cdots \cdots \cdots \\ a_{l0} \ a_{l1} \ \cdots \ a_{lk} \end{bmatrix},$$

where $a_{ij} = w(t_i, p_j) - w(p_j, t_i)$, $i = 1, 2, \ldots, l$, and $j = 1, 2, \ldots, k$.

The next marking $\mathbf{m}'$ of MPN is calculated according to the so-called vector state equation, i.e.:

$$\mathbf{m}' = \mathbf{m} + \mathbf{f}\mathbf{A}.$$

There are many different classes of Petri nets extending the basic definition. In [13], the Petri nets were extended with inhibitor arcs. The inhibitor arcs test the absence of tokens in places and they can be used to disable transitions. A transition can only fire if all its places connected through inhibitor arcs are empty.

Definition 4.4 A marked Petri net with inhibitor arcs is a five-tuple $MPN^\circ = (Pl, Tr, Arc, w, m)$, where:

- Pl is the finite set of places,
- Tr is the finite set of transitions,
- $Arc = Arc_O \cup Arc_I$ such that $Arc_O \subseteq (Pl \times Tr) \cup (Tr \times Pl)$ is the set of ordinary arcs from places to transitions and from transitions to places whereas $Arc_I \subseteq (Pl \times Tr)$ is the set of inhibitor arcs from places to transitions,
- $w : A \rightarrow \{1, 2, 3, \ldots\}$ is the weight function on the arcs,
- $m_0 : Pl \rightarrow \{0, 1, 2, \ldots\}$ is the initial marking function on the places.

For the marked Petri net $MPN^\circ = (Pl, Tr, Arc, w, m)$ with inhibitor arcs, we define for any $t \in Tr$:

- $I_O(t) = \{p \in Pl : (p, t) \in Arc_O\}$—a set of input places connected through ordinary arcs to the transition t,
- $I_I(t) = \{p \in Pl : (p, t) \in Arc_I\}$—a set of input places connected through inhibitor arcs to the transition t,
- $O(t) = \{p \in Pl : (t, p) \in Arc_O\}$—a set of output places connected through ordinary arcs from the transition t.

Analogously to marked Petri nets, we define, for the marked Petri net $MPN^\circ = (Pl, Tr, Arc, w, m)$ with inhibitor arcs, a mapping $m : Pl \rightarrow \{0, 1, 2, \ldots\}$ assigning tokens to places.

Definition 4.5 Let $MPN^\circ = (Pl, Tr, Arc, w, m_0)$ be a marked Petri net with inhibitor arcs. A transition $t \in Tr$ is said to be enabled if and only if $m(p) \geq w(p, t)$ for all $p \in I_O(t)$ and $m(p) = 0$ for all $p \in I_I(t)$.

The vector state equation for calculating the next marking $\mathbf{m}'$ of MPN° is analogous to that for the marked Petri net. The difference lies only in the fact that, in the incidence matrix $\mathbf{A}$, only weights of ordinary arcs are taken into consideration.

4.2 Petri Net Models of Logic Gates for Physarum Machines

Petri nets with inhibitor arcs can be used to model plasmodium propagation in *Physarum* machines used to simulate combinational logic circuits (see [119]).

Let a *Physarum* machine $\mathscr{PM}$ simulate a combinational logic circuit. In the proposed Petri net models, we can distinguish three kinds of places:

- Places representing the original points of plasmodium of *Physarum polycephalum* in a *Physarum* machine $\mathscr{PM}$. For each place of this type, $m(p) = 1$ means that plasmodium of *Physarum polycephalum* is present in the original point represented by p. Otherwise, $m(p) = 0$ means that there is no plasmodium in the original point.
- Places associated with inputs of a combinational logic circuit and representing attractants / repellents in a *Physarum* machine $\mathscr{PM}$. For each place of this type, $m(p) = 1$ means that the associated input is equal to 1 and the attractant / repellent is activated. Otherwise, $m(p) = 0$ means that the associated input is equal to 0 and the attractant / repellent is deactivated.
- Places representing attractants in a *Physarum* machine $\mathscr{PM}$ determining outputs of a combinational logic circuit. For each place of this type, $m(p) = 1$ denotes the presence of plasmodium of *Physarum polycephalum* at the attractant (plasmodium of *Physarum polycephalum* occupies the attractant). It means that the output is equal to 1. Otherwise, $m(p) = 0$ denotes the absence of plasmodium of *Physarum polycephalum* at the attractant. It means that the output is equal to 0.

Additionally, the following limits are assumed for the Petri net modeling plasmodium propagation:

- $w(a) = 1$ for each $a \in Arc$,
- $m(p) \leq 1$ for each $p \in Pl$ (the capacity limit).

Let $MPN^\circ = (Pl, Tr, Arc, w, m_0)$ be a marked Petri net with inhibitor arcs. Assuming limits as above, a transition $t \in Tr$ is said to be enabled if and only if $m(p) = 1$ for all $p \in I_O(t)$ and $m(p) = 0$ for all $p \in I_I(t)$ and $m(p) = 0$ for all $p \in O(t)$.

Remark 4.1 In all Petri net models shown in figures, to simplify them, we have used bidirectional ordinary arcs between input places and transitions instead of ordinary arcs from input places to transitions and from transitions to input places. A bidirectional arc causes that the token is not consumed (removed) from the input place after firing a transition. This fact has a natural justification, i.e., plasmodium propagation does not cause that the attractants are deactivated and plasmodium disappears from the original points.

The distribution of stimuli in the *Physarum* machine $\mathscr{PM}_{AND}$ simulating the AND gate is shown in Fig. 4.1.

Fig. 4.1 The distribution of stimuli in the *Physarum* machine $\mathscr{PM}_{AND}$

Fig. 4.2 The Petri net model of the AND gate for the *Physarum* machine

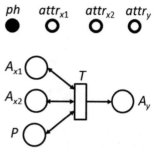

For the geometrical distribution of attractants, we assume that:

- *ph* belongs to $ROI(attr_{x1})$,
- $attr_{x1}$ belongs to $ROI(attr_{x2})$,
- $attr_{x2}$ belongs to $ROI(attr_y)$,

where *ROI* is the region of influence. One can see that the propagation of plasmodium from the original point *ph* to the output attractant $attr_y$ determining the output of the AND gate is possible if and only if both attractants $attr_{x_1}$ and $attr_{x_2}$ are activated. We assume that the attractant $attr_y$ is always activated.

The Petri net model of the AND gate for the *Physarum* machine is shown in Fig. 4.2. In this figure:

- A_{x_1} is a place associated with the input x_1 and representing the attractant $attr_{x1}$.
- A_{x_2} is a place associated with the input x_2 and representing the attractant $attr_{x2}$.
- A_y is a place representing the attractant $attr_y$ determining the output *y*.
- *P* is a place representing the original point of plasmodium.

The transition *T* represents the flow (propagation) of plasmodium from the original point *ph* to the attractant $attr_y$. *T* is enabled to fire if, among others, both $m(A_{x1}) = 1$ and $m(A_{x2}) = 1$, i.e. both attractants, $attr_{x1}$ and $attr_{x2}$, are activated.

In the Petri net model of the AND gate, we have $Pl = \{A_{x_1}, A_{x_2}, A_y, P\}$ and $Tr = \{T\}$. The incidence matrix has the form:

$$A = \begin{bmatrix} 0 \\ 0 \\ 0 \\ 1 \end{bmatrix}.$$

Figures 4.3, 4.4, 4.5, and 4.6 show the behaviour of the Petri net model of the AND gate for all possible combinations of input values of x_1 and x_2.

The distribution of stimuli in the *Physarum* machine $\mathscr{PM}_{OR}$ simulating the OR gate is shown in Fig. 4.7.

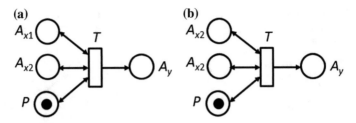

Fig. 4.3 The Petri net model behaviour of the AND gate for $x_1 = 0$ and $x_2 = 0$, **a** before transition firing, **b** after transition firing

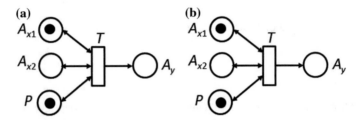

Fig. 4.4 The Petri net model behaviour of the AND gate for $x_1 = 1$ and $x_2 = 0$, **a** before transition firing, **b** after transition firing

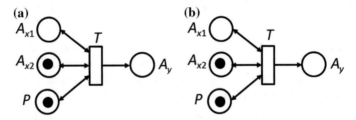

Fig. 4.5 The Petri net model behaviour of the AND gate for $x_1 = 0$ and $x_2 = 1$, **a** before transition firing, **b** after transition firing

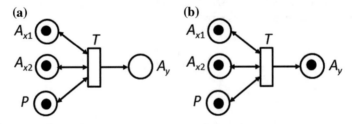

Fig. 4.6 The Petri net model behaviour of the AND gate for $x_1 = 1$ and $x_2 = 1$, **a** before transition firing, **b** after transition firing

Fig. 4.7 The distribution of stimuli in the *Physarum* machine $\mathscr{P}\mathscr{M}_{OR}$

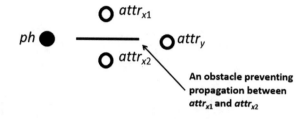

Fig. 4.8 The Petri net model of the OR gate for the *Physarum* machine

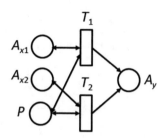

For the geometrical distribution of attractants, we assume that:

- *ph* belongs to $ROI(attr_{x1})$ and $ROI(attr_{x2})$,
- $attr_{x1}$ belongs to $ROI(attr_y)$,
- $attr_{x2}$ belongs to $ROI(attr_y)$,

where *ROI* is the region of influence. One can see that the propagation of plasmodium from the original point *ph* to the output attractant $attr_y$ determining the output of the OR gate is possible if and only if at least one of the attractants ($attr_{x1}$ or $attr_{x2}$) is activated. We assume that the attractant $attr_y$ is always activated.

The Petri net model of the OR gate for the *Physarum* machine is shown in Fig. 4.8. The meaning of elements (excluding transitions) in this figure is analogous to those described for the AND gate. The transitions T_1 and T_2 represent the alternative flows of plasmodium from the original point *ph* to the attractant $attr_y$. T_1 is enabled to fire if, among others, $m(A_{x1}) = 1$, i.e. the attractant $attr_{x1}$ is activated. T_2 is enabled to fire if, among others, $m(A_{x2}) = 1$, i.e. the attractant $attr_{x2}$ is activated.

In the Petri net model of the OR gate, we have $Pl = \{A_{x_1}, A_{x_2}, A_y, P\}$ and $Tr = \{T_1, T_2\}$. The incidence matrix has the form:

$$\mathbf{A} = \begin{bmatrix} 0 & 0 \\ 0 & 0 \\ 0 & 0 \\ 1 & 0 \end{bmatrix}.$$

Figures 4.9, 4.10, 4.11, and 4.12 show the behaviour of the Petri net model of the OR gate for all possible combinations of input values of x_1 and x_2.

The distribution of stimuli in the *Physarum* machine $\mathscr{P}\mathscr{M}_{NOT}$ simulating the NOT gate is shown in Fig. 4.13.

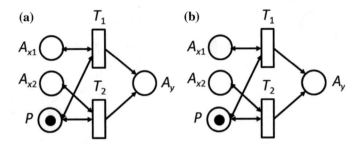

Fig. 4.9 The Petri net model behaviour of the OR gate for $x_1 = 0$ and $x_2 = 0$, **a** before transition firing, **b** after transition firing

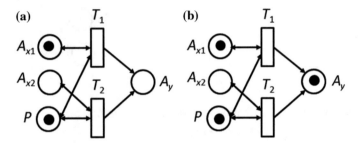

Fig. 4.10 The Petri net model behaviour of the OR gate for $x_1 = 1$ and $x_2 = 0$, **a** before transition firing, **b** after transition firing

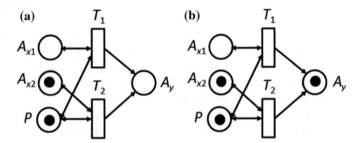

Fig. 4.11 The Petri net model behaviour of the OR gate for $x_1 = 0$ and $x_2 = 1$, **a** before transition firing, **b** after transition firing

For the geometrical distribution of attractants, we assume that:

- *ph* belongs to *ROI*(*attr*$_a$),
- *attr*$_a$ belongs to *ROI*(*attr*$_y$),

where *ROI* is the region of influence. One can see that the propagation of plasmodium from the original point *ph* to the attractant *attr*$_y$ determining the output of the NOT gate is possible if and only if the repellent *rep*$_x$ is deactivated. Otherwise, if *rep*$_x$ is activated, it avoids propagation of plasmodium toward the auxiliary attractant *attr*$_a$,

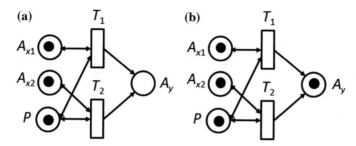

Fig. 4.12 The Petri net model behaviour of the OR gate for $x_1 = 1$ and $x_2 = 1$, **a** before transition firing, **b** after transition firing

Fig. 4.13 The distribution of stimuli in the *Physarum* machine $\mathscr{PM}_{NOT}$

Fig. 4.14 The Petri net model of the NOT gate for the *Physarum* machine

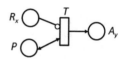

and next, toward the output attractant $attr_y$. We assume that the attractant $attr_y$ is always activated.

The Petri net model of the NOT gate for the *Physarum* machine is shown in Fig. 4.14. In this figure:

- R_x is a place associated with the input x and representing the repellent rep_x.
- A_y is a place representing the attractant $attr_y$ determining the output y.
- P is a place representing the original point of plasmodium.

In the Petri net model of the NOT gate, we have $Pl = \{R_x, A_y, P\}$ and $Tr = \{T\}$. The incidence matrix has the form:

$$A = \begin{bmatrix} 0 \\ 0 \\ 1 \end{bmatrix}.$$

The transition T represents the flow (propagation) of plasmodium from the original point ph to the attractant $attr_y$. T is enabled to fire if, among others, $m(R_x) = 0$, i.e. the repellent rep_x is deactivated.

Figures 4.15 and 4.16 show the behaviour of the Petri net model of the NOT gate for all possible combinations of input values of x.

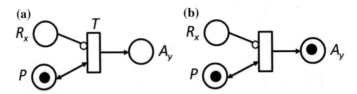

Fig. 4.15 The Petri net model behaviour of the NOT gate for $x = 0$, **a** before transition firing, **b** after transition firing

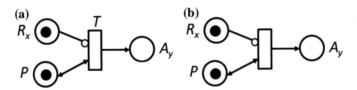

Fig. 4.16 The Petri net model behaviour of the NOT gate for $x = 1$, **a** before transition firing, **b** after transition firing

Fig. 4.17 A schematic symbol of the 2-to-1 multiplexer

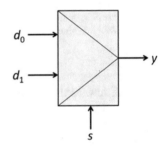

4.3   A Petri Net Model of a Multiplexer for Physarum Machines

A multiplexer is a device that selects one of the several inputs and forwards the selected input into a single output line. In Fig. 4.17, a schematic symbol of the 2-to-1 multiplexer is shown.

In the schematic symbol: d_0, d_1 are data inputs, s is a selection input, and y is an output. The operation of the multiplexer can be described as follows:

- if $s = 0$, then $y = d_0$,
- if $s = 1$, then $y = d_1$.

The functional specification can be written as $y = \bar{s}d_0 + sd_1$ (Fig. 4.18).

Figures 4.19, 4.20, 4.21, 4.22, 4.23, 4.24, 4.25, and 4.26 show the behaviour of the Petri net model of the 2-to-1 multiplexer for all possible combinations of input values of s, d_0, and d_1.

The Petri net model (Fig. 4.18) can be translated into the lower level language, i.e., the geometrical distribution of attractants and repellents that is shown in Fig. 4.27,

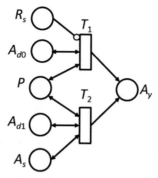

Fig. 4.18   The Petri net model of the 2-to-1 multiplexer for the *Physarum* machine

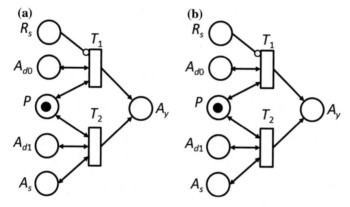

Fig. 4.19   The Petri net model behaviour of the 2-to-1 multiplexer for $s = 0$, $d_0 = 0$, and $d_1 = 0$, **a** before transition firing, **b** after transition firing

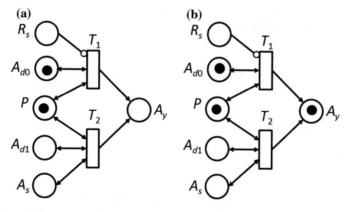

Fig. 4.20   The Petri net model behaviour of the 2-to-1 multiplexer for $s = 0$, $d_0 = 1$, and $d_1 = 0$, **a** before transition firing, **b** after transition firing

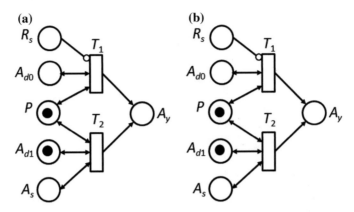

Fig. 4.21 The Petri net model behaviour of the 2-to-1 multiplexer for $s = 0$, $d_0 = 0$, and $d_1 = 1$, **a** before transition firing, **b** after transition firing

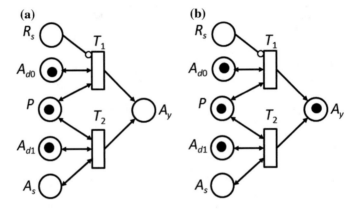

Fig. 4.22 The Petri net model behaviour of the 2-to-1 multiplexer for $s = 0$, $d_0 = 1$, and $d_1 = 1$, **a** before transition firing, **b** after transition firing

where ph denotes the original point of plasmodium of *Physarum polycephalum*, $attr_{d0}, attr_{d1}, attr_s, attr_y$—the attractants, and rep_s—the repellent. For the geometrical distribution of stimuli, we assume that:

- ph belongs to $ROI(attr_{d0})$, $ROI(rep_s)$, and $ROI(attr_s)$,
- $attr_s$ belongs to $ROI(attr_{d1})$,
- $attr_{d0}$ belongs to $ROI(attr_y)$,
- $attr_{d1}$ belongs to $ROI(attr_y)$,

where ROI is the region of influence. Logic states are implemented in the following way:

- $s = 0$ means R_s and A_s are deactivated, $s = 1$ means R_s and A_s are activated,
- $d_0 = 0$ means A_{d0} is deactivated, $d_0 = 1$ means A_{d0} is activated,

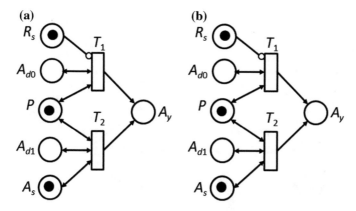

Fig. 4.23 The Petri net model behaviour of the 2-to-1 multiplexer for $s = 1$, $d_0 = 0$, and $d_1 = 0$, **a** before transition firing, **b** after transition firing

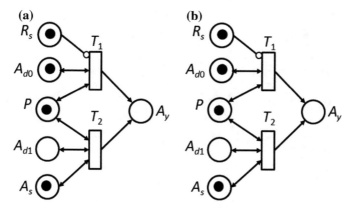

Fig. 4.24 The Petri net model behaviour of the 2-to-1 multiplexer for $s = 1$, $d_0 = 1$, and $d_1 = 0$, **a** before transition firing, **b** after transition firing

- $d_1 = 0$ means A_{d1} is deactivated, $d_1 = 1$ means A_{d1} is activated.

It is worth noting that A_y is always activated.

4.4   A Petri Net model of a Demultiplexer for Physarum Machines

A demultiplexer is a device taking a single input signal and selecting one of many data-output-lines, which is connected to the single input. In Fig. 4.28, a schematic symbol of the 1-to-2 demultiplexer is shown (Fig. 4.29). In the schematic symbol:

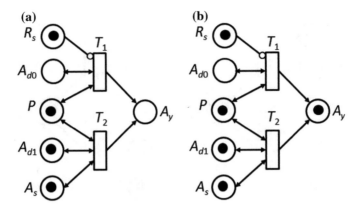

Fig. 4.25 The Petri net model behaviour of the 2-to-1 multiplexer for $s = 1$, $d_0 = 0$, and $d_1 = 1$, **a** before transition firing, **b** after transition firing

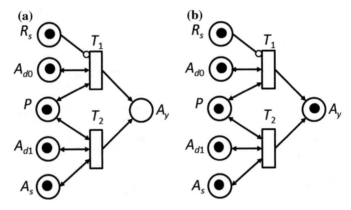

Fig. 4.26 The Petri net model behaviour of the 2-to-1 multiplexer for $s = 1$, $d_0 = 1$, and $d_1 = 1$, **a** before transition firing, **b** after transition firing

Fig. 4.27 The geometrical distribution of attractants and repellents for the 2-to-1 multiplexer

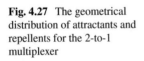

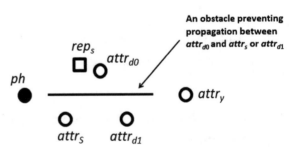

Fig. 4.28 A schematic symbol of the 1-to-2 demultiplexer

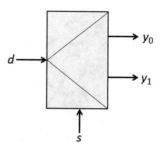

Fig. 4.29 The Petri net model of the 1-to-2 demultiplexer for the *Physarum* machine

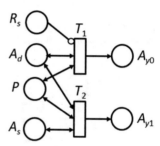

d is a data input, s is a selection input, and y_0, y_1 are outputs. The operation of the demultiplexer can be described as follows:

- if $s = 0$, then $y_0 = d$,
- if $s = 1$, then $y_1 = d$.

The functional specification can be written as $y_0 = \bar{s}d$ and $y_1 = sd$.

Figures 4.30, 4.31, 4.32, and 4.33 show the behaviour of the Petri net model of the 1-to-2 demultiplexer for all possible combinations of input values of s and d.

The Petri net model (Fig. 4.29) can be translated into the lower level language, i.e., the geometrical distribution of attractants and repellents that is shown in Fig. 4.34, where ph denotes the original point of plasmodium of *Physarum polycephalum*, $attr_d$, $attr_s$, $attr_{y0}$, $attr_{y1}$ are the attractants, and rep_s is the repellent. For the geometrical distribution of stimuli, we assume that:

- ph belongs to $ROI(attr_d)$,
- $attr_d$ belongs to $ROI(attr_{y0})$, $ROI(rep_s)$, and $ROI(attr_s)$,
- $attr_s$ belongs to $ROI(attr_{y1})$,

where ROI is the region of influence. Logic states are implemented in the following way:

- $s = 0$ means R_s and A_s are deactivated, $s = 1$ means R_s and A_s are activated,
- $d = 0$ means A_d is deactivated, $d = 1$ means A_d is activated.

It is worth noting that A_{y0} and A_{y1} are always activated.

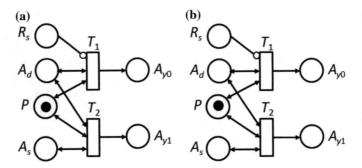

Fig. 4.30 The Petri net model behaviour of the 1-to-2 demultiplexer for $s = 0$ and $d = 0$, **a** before transition firing, **b** after transition firing

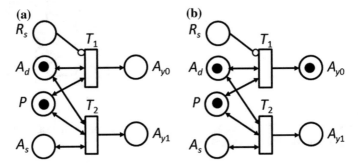

Fig. 4.31 The Petri net model behaviour of the 1-to-2 demultiplexer for $s = 0$ and $d = 1$, **a** before transition firing, **b** after transition firing

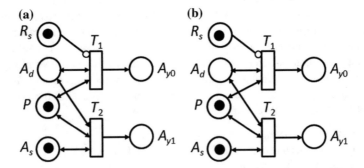

Fig. 4.32 The Petri net model behaviour of the 1-to-2 demultiplexer for $s = 1$ and $d = 0$, **a** before transition firing, **b** after transition firing

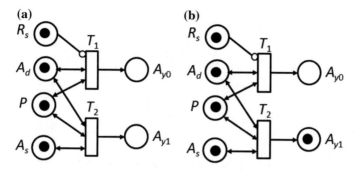

Fig. 4.33 The Petri net model behaviour of the 1-to-2 demultiplexer for $s = 1$ and $d = 1$, **a** before transition firing, **b** after transition firing

Fig. 4.34 The geometrical distribution of attractants and repellents for the 1-to-2 demultiplexer

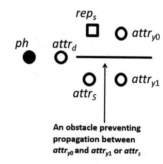

4.5   A Petri Net Model of a Half Adder for Physarum Machines

A half adder adds two single binary digits a and b. It has two outputs, sum and carry. The carry signal represents an overflow into the next digit of a multi-digit addition. In Fig. 4.35, a schematic symbol of the half adder is shown (Fig. 4.36).

In the schematic symbol: a, b are one-bit inputs, s is a sum output, and c is a carry output. The functional specification of the half adder can be written as:

- $s = \overline{a}b + a\overline{b}$,
- $c = ab$.

Fig. 4.35 A schematic symbol of the half adder

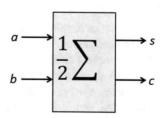

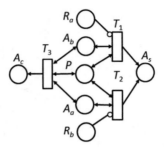

Fig. 4.36 The Petri net model of the half adder for the *Physarum* machine

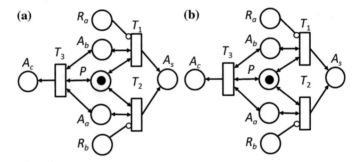

Fig. 4.37 The Petri net model behaviour of the half adder for $a = 0$ and $b = 0$, **a** before transition firing, **b** after transition firing

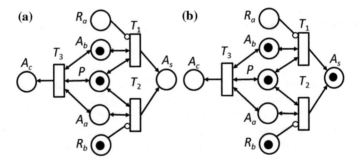

Fig. 4.38 The Petri net model behaviour of the half adder for $a = 1$ and $b = 0$, **a** before transition firing, **b** after transition firing

Figures 4.37, 4.38, 4.39, and 4.40 show the behaviour of the Petri net model of the half adder for all possible combinations of input values of a and b.

The Petri net model (Fig. 4.36) can be translated into the lower level language, i.e., the geometrical distribution of attractants and repellents that is shown in Fig. 4.41, where *ph* denotes the original point of plasmodium of *Physarum polycephalum*, $attr_a$, $attr_b$, $attr_s$, $attr_c$ are the attractants, and rep_a, rep_b are the repellents. For the geometrical distribution of stimuli, we assume that:

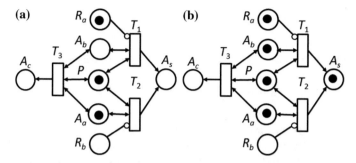

Fig. 4.39 The Petri net model behaviour of the half adder for $a = 0$ and $b = 1$, **a** before transition firing, **b** after transition firing

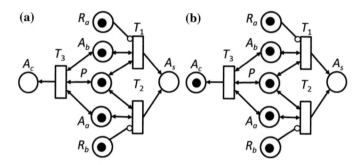

Fig. 4.40 The Petri net model behaviour of the half adder for $a = 1$ and $b = 1$, **a** before transition firing, **b** after transition firing

Fig. 4.41 The geometrical distribution of attractants and repellents for the half adder

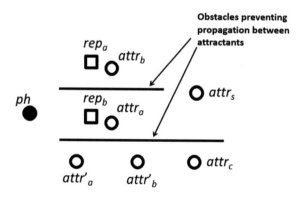

- ph belongs to $ROI(attr_a)$, $ROI(attr_b)$, $ROI(rep_a)$, and $ROI(rep_b)$,
- $attr_a$ and $attr_b$ belong to $ROI(attr_s)$,
- $attr'_a$ belongs to $ROI(attr'_b)$,
- $attr'_b$ belongs to $ROI(attr_c)$,

where *ROI* is the region of influence. Logic states are implemented in the following way:

- $a = 0$ means A_a, A'_a, and R_a are deactivated, $a = 1$ means A_a, A'_a, and R_a are activated,
- $b = 0$ means A_b, A'_b, and R_b are deactivated, $b = 1$ means A_b, A'_b, and R_b are activated.

It is worth noting that A_s and A_c are always activated.

As we have shown, the slime mould can implement the Petri nets. This means that the slime mould can be used as a kind of simple computer that can calculate and solve some simple tasks.

Chapter 5
Rough Set Based Descriptions of Plasmodium Propagation

5.1 The Rudiments of Rough Sets

The main problem with formalizing the slime mould behaviours is that *Physarum polycephalum* tries to occupy all achievable goals simultaneously. In other words, timed transition systems realized by the slime mould cannot be monogenic. For these special cases, we can appeal to rough sets defined on transitions, see [58, 63, 122, 123, 126]. Let us remind that rough sets theory proposed by Z. Pawlak is a mathematical approach to imperfect knowledge [64–66]. Rough sets are an appropriate tool to deal with ambigous or imprecise concepts in the universe of discourse.

The idea of rough sets consists of the approximation of a given set by a pair of sets, called the lower and upper approximation of this set. Some sets cannot be exactly defined. If a given set X is not exactly defined, then we employ two exact sets (the lower and the upper approximation of X) that define X roughly (approximately).

In general, a rough set is defined on the basis of any binary relation between objects in some universe of discourse (see [150]).

Let $U \neq \emptyset$ be a finite set of objects we are interested in. U is called the universe. Let R be any binary relation over U. With each subset $X \subseteq U$ and any binary relation R over U, we associate two subsets:

- $\underline{R}(X) = \{u \in U : R(u) \subseteq X\}$,
- $\overline{R}(X) = \{u \in U : R(u) \cap X \neq \emptyset\}$,

called the R-lower and R-upper approximation of X, respectively. A set $BN_R(X) = \overline{R}(X) - \underline{R}(X)$ is called the R-boundary region of X. If $BN_R(X) = \emptyset$, then X is sharp (exact) with respect to R. Otherwise, X is rough (inexact).

Roughness of a set X can be characterized numerically. To this end, the accuracy of approximation of X with respect to R is defined as:

© Springer International Publishing AG, part of Springer Nature 2019
A. Schumann and K. Pancerz, *High-Level Models of Unconventional Computations*, Studies in Systems, Decision and Control 159,
https://doi.org/10.1007/978-3-319-91773-3_5

$$\alpha_R(X) = \frac{card\,(\underline{R}(X))}{card\,(\overline{R}(X))},$$

where *card* denotes the cardinality of the set and $X \neq \emptyset$.

The definitions given earlier are based on the standard definition of set inclusion. Let U be the universe and $A, B \subseteq U$. The standard set inclusion is defined as

$$A \subseteq B \text{ if and only if } \underset{u \in A}{\forall}\; u \in B.$$

In some situations, the application of this definition seems to be too restrictive and rigorous. W. Ziarko proposed in [151] some relaxation of the original rough set approach. His proposition is called the Variable Precision Rough Set Model (VPRSM). The VPRSM approach is based on the notion of majority set inclusion. Let U be the universe, $A, B \subseteq U$, and $0 \leq \beta < 0.5$. The majority set inclusion is defined as

$$A \overset{\beta}{\subseteq} B \text{ if and only if } 1 - \frac{card\,(A \cap B)}{card\,(A)} \leq \beta,$$

where *card* denotes the cardinality of the set. $A \overset{\beta}{\subseteq} B$ means that a specified majority of elements belonging to A belongs also to B. One can see that if $\beta = 0$, then the majority set inclusion becomes the standard set inclusion.

By replacing the standard set inclusion with the majority set inclusion in definitions of approximations, we obtain the following two subsets:

- $\underline{R}^{\beta}(X) = \{u \in U : R(u) \overset{\beta}{\subseteq} X\}$,
- $\overline{R}^{\beta}(X) = \{u \in U : 1 - \frac{card\,(R(u) \cap X)}{card\,(R(u))} < 1 - \beta\}$,

called the R_{β}-lower and R_{β}-upper approximation of X, respectively.

The approach presented in this chapter refers to a general framework for the study of approximation using the notion of neighborhood systems proposed by T. Y. Lin (cf. [51]).

5.2 Rough Set Descriptions Based on Transition Systems

As we already know, the plasmodium propagation in *Physarum* machines can be well modelled by means of transition systems within a process algebra or spatial logic. In addition, some descriptions of plasmodium propagation can be considered in terms of rough set theory (see [61]). The point is that in the behaviour of *Physarum* machines, one can notice some ambiguities in plasmodium motions, which make a cloud anticipation of states of machines in time. To identify these ambiguities, rough set descriptions created over transition systems were proposed in [61].

If a given transition system $TS(\mathcal{PM}) = (S, E, T, S_{init})$ modeling the plasmodium propagation in the *Physarum* machine $\mathcal{PM}$ is not monogenic, one can see that we deal with ambiguities of direct successors of some states, i.e., there exist states having no uniquely determined direct successors. We can manage these ambiguities using rough set theory. Analogously to approximation of sets defined in rough set theory, we can define rough anticipation of states over transition systems. The anticipation of states is made via direct predecessor states of the ones anticipated. Therefore, we call this anticipation the predecessor anticipation.

Definition 5.1 Let $TS(\mathscr{PM}) = (S, E, T, S_{init})$ be a transition system modeling the plasmodium propagation in a given *Physarum* machine $\mathscr{PM}$ and $X \subseteq S$ be a set of distinguished states. The lower predecessor anticipation $\underline{Pre}(X)$ of X is defined as

$$\underline{Pre}(X) = \{s \in S : Post(s) \neq \emptyset \text{ and } Post(s) \subseteq X\}.$$

The upper predecessor anticipation $\overline{Pre}(X)$ of X is defined as

$$\overline{Pre}(X) = \{s \in S : Post(s) \cap X \neq \emptyset\}.$$

The lower predecessor anticipation consists of all states from which TS surely goes immediately to the distinguished states included in X as results of any events occurring at these states. The upper predecessor anticipation consists of all states from which TS possibly goes immediately to the distinguished states included in X as results of some events occurring at these states. It means that TS can also go immediately to the states from outside X.

The set $BN_{Pre}(X) = \overline{Pre}(X) - \underline{Pre}(X)$ is referred to as the boundary region of predecessor anticipation of X.

Specifically, we can determine the lower and upper predecessor anticipation of a single state $s \in S$, denoted as $\underline{Pre}(s)$ and $\overline{Pre}(s)$, respectively. Moreover, the boundary region of predecessor anticipation of s is denoted as $BN_{Pre}(s)$.

If $BN_{Pre}(X) = \emptyset$, then the anticipation of the set X of states on the basis of their direct predecessors is sharp. In the opposite case (i.e. $BN_{Pre}(X) \neq \emptyset$), the anticipation of X is rough. The accuracy of anticipation can be defined analogously to the accuracy of approximation in rough set theory, i.e.:

$$\alpha(X) = \frac{\underline{Pre}(X)}{\overline{Pre}(X)}.$$

Example 5.1 Let us consider a transition system $TS(\mathscr{PM}) = (S, E, T, S_{init})$, from Example 3.5, modeling the plasmodium propagation in a given *Physarum* machine $\mathscr{PM}$. Let us be interested in the following set of distinguished states $X = \{s_5, s_6, s_8\}$. For X, we obtain:

- $\underline{Pre}(X) = \{s_3\}$ because $Post(s_3) \neq \emptyset$ and $Post(s_3) \subseteq X$,
- $\overline{Pre}(X) = \{s_2, s_3\}$ because $Post(s_2) \cap X \neq \emptyset$ and $Post(s_3) \cap X \neq \emptyset$.

Hence, $BN_{Pre}(X) \neq \emptyset$. It means that the anticipation of states from X is rough on the basis of their direct predecessors. The accuracy of anticipation $\alpha(X)$ is equal to $\frac{1}{2}$.

Using the rough set description, we can determine, for each state $s \in S$ in a transition system $TS(\mathscr{P}\mathscr{M}) = (S, E, T, S_{init})$, modeling the plasmodium propagation in a given *Physarum* machine $\mathscr{P}\mathscr{M}$, the core originators of s.

Definition 5.2 Let $TS(\mathscr{P}\mathscr{M}) = (S, E, T, S_{init})$ be a transition system modeling the plasmodium propagation in a given *Physarum* machine $\mathscr{P}\mathscr{M}$. Any initial state $s' \in S_{init}$ of a given state $s \in S$ is called the core originator of s if and only if for each possible string $v = e_1 e_2 \ldots e_{k-1}$ of events in TS such that $s_1 \xrightarrow{e_1} s_2 \xrightarrow{e_2} \ldots \xrightarrow{e_{k-1}} s_k$ and $Post(s_k) = \emptyset$ (i.e. s_k is a goal state in TS):

$$s \in \{s_1, s_2, \ldots, s_k\}.$$

In case of core originators of a given state s, all predecessors (both direct and indirect) of s explicitly determine the way leading to s. The set of all core originators of s is denoted by $CoreOrig(s)$.

To determine the set of all core originators of a given state $s \in S$ in a transition system $TS(\mathscr{P}\mathscr{M}) = (S, E, T, S_{init})$ modeling the plasmodium propagation in a given *Physarum* machine $\mathscr{P}\mathscr{M}$, we may use Algorithm 1 (cf. [61]).

Algorithm 1: Algorithm for searching for the set of all core originators of a given state s.

Input : $TS(\mathscr{P}\mathscr{M}) = (S, E, T, S_{init})$ is a transition system modeling plasmodium propagation in a given *Physarum* machine $\mathscr{P}\mathscr{M}$, $s \in S$.
Output: $CoreOrig(s)$ is the set of all core originators of s.
$CoreOrig(s) \leftarrow \emptyset$;
$Y \leftarrow \{s\}$;
while $Y \neq \emptyset$ **do**
    $Z \leftarrow \underline{Pre}(Y)$;
    $Y \leftarrow Z$;
    for *each* $s' \in Z$ **do**
        if $s' \in S_{init}$ **then**
            $CoreOrig(s) \leftarrow CoreOrig(s) \cup \{s'\}$;
            $Y \leftarrow Y - \{s'\}$;

Return $CoreOrig(s)$;

Example 5.2 Let us consider a transition system $TS = (S, E, T, S_{init})$, from Example 3.5, modeling the plasmodium propagation in a given *Physarum* machine $\mathscr{P}\mathscr{M}$. The state s_1 is a core originator of the states s_3, s_4, and s_8.

We slightly modify definition of the lower and upper predecessor anticipation in case of the Variable Precision Rough Set Model (VPRSM).

Definition 5.3 Let $TS(\mathscr{PM}) = (S, E, T, S_{init})$ be a transition system modeling the plasmodium propagation in a given *Physarum* machine $\mathscr{PM}$, $X \subseteq S$ be a set of distinguished states, and $0 \leq \beta < 0.5$. The β-lower predecessor anticipation $\underline{Pre}^{\beta}(X)$ of X is defined as

$$\underline{Pre}^{\beta}(X) = \{s \in S : Post(s) \neq \emptyset \text{ and } Post(s) \overset{\beta}{\subseteq} X\}.$$

The β-upper predecessor anticipation $\overline{Pre}^{\beta}(X)$ of X is defined as

$$\overline{Pre}^{\beta}(X) = \{s \in S : 1 - \frac{Post(s) \cap X}{Post(s)} < 1 - \beta\}.$$

Now, one can see that the β-lower predecessor anticipation consists of each state from which TS goes immediately, in most cases (i.e. in terms of the majority set inclusion), to the distinguished states included in X as results of some events occurring at these states.

As before, specifically, we can determine the β-lower and β-upper predecessor anticipation of a single state $s \in S$, denoted as $\underline{Pre}^{\beta}(s)$ and $\overline{Pre}^{\beta}(s)$, respectively. Moreover, the β-boundary region of predecessor anticipation of s is denoted as $BN_{Pre}^{\beta}(s)$.

Example 5.3 Let us consider a transition system $TS(\mathscr{PM}) = (S, E, T, S_{init})$, from Example 3.5, modeling the plasmodium propagation in a given *Physarum* machine $\mathscr{PM}$. Let us be interested in the following set of distinguished states $X = \{s_5, s_6, s_8\}$. For X and $\beta = 0.5$, we obtain:

- $\underline{Pre}^{0.5}(X) = \{s_2, s_3\}$ because $Post(s_2) \neq \emptyset, Post(s_2) \overset{0.5}{\subseteq} X, Post(s_3) \neq \emptyset$, and $Post(s_3) \overset{0.5}{\subseteq} X$,
- $\overline{Pre}^{0.5}(X) = \{s_2, s_3\}$.

Definition 5.4 Let $TS(\mathscr{PM}) = (S, E, T, S_{init})$ be a transition system modeling the plasmodium propagation in a given *Physarum* machine $\mathscr{PM}$, $X \subseteq S$, and $0 \leq \beta < 0.5$. The state s' is said to be:

- a strict anticipator of the states from X if and only if $s \in \underline{Pre}(X)$ (i.e. the case when $\beta = 0$),
- a quasi-strict anticipator of the states from X if and only if $s \in \underline{Pre}^{\beta}(X)$ only for $\beta > 0$.

Example 5.4 Let us consider a transition system $TS(\mathscr{PM}) = (S, E, T, S_{init})$, from Example 3.5, modeling the plasmodium propagation in a given *Physarum* machine $\mathscr{PM}$. For $X = \{s_5, s_6, s_8\}$ and $\beta = 0.5$:

- s_3 is a strict anticipator of the states from X because $s_3 \in \underline{Pre}(X)$,
- s_2 is a quasi-strict anticipator of the states from X because $s_2 \in \underline{Pre}^{0.5}(X)$, but $s_2 \notin \underline{Pre}(X)$.

We slightly modify definition of the lower and upper predecessor anticipation in case of models in the form of timed transition systems. Now, anticipations are defined for each time instant t because the lower and upper predecessor anticipation can change over time.

Definition 5.5 Let $TTS(\mathscr{PM}) = (S, E, T, S_{init}, l, u)$ be a timed transition system modeling the plasmodium propagation in a given *Physarum* machine $\mathscr{PM}$, $X \subseteq S$ be a set of distinguished states. The lower predecessor anticipation $\underline{Pre}_t(X)$ of X at the time instant t is defined as

$$\underline{Pre}_t(X) = \{s \in S : Post_t(s) \neq \emptyset \text{ and } Post_t(s) \subseteq X\}.$$

The upper predecessor anticipation $\overline{Pre}_t(X)$ of X at the time instant t is defined as

$$\overline{Pre}_t(X) = \{s \in S : Post_t(s) \cap X \neq \emptyset\}.$$

The lower predecessor anticipation $\underline{Pre}_t(X)$ consists of all states from which TTS surely goes immediately to the distinguished states included in X as results of any events occurring at these states at the time instant t. The upper predecessor anticipation $\overline{Pre}_t(X)$ consists of all states from which TTS possibly goes immediately to the distinguished states included in X as results of some events occurring at these states at the time instant t.

Example 5.5 Let us consider a timed transition system $TTS(\mathscr{PM}) = (S, E, T, S_{init}, l, u)$, from Example 3.7, modeling the plasmodium propagation in a given *Physarum* machine $\mathscr{PM}$. Let us be interested in the following set of distinguished states $X = \{s_5, s_6, s_8\}$. For X, we obtain:

- $\underline{Pre}_t(X) = \{s_3\}$ if $t < 1$ or $t > 2$ because $Post_t(s_3) \neq \emptyset$ and $Post_t(s_3) \subseteq X$,
- $\overline{Pre}_t(X) = \{s_2, s_3\}$ if $t < 1$ or $t > 2$ because $Post_t(s_2) \cap X \neq \emptyset$ and $Post_t(s_3) \cap X \neq \emptyset$,
- $\underline{Pre}_t(X) = \{s_2, s_3\}$ if $t \geq 1$ and $t \leq 2$ because $Post_t(s_2) \neq \emptyset$, $Post_t(s_2) \subseteq X$, $Post_t(s_3) \neq \emptyset$, and $Post_t(s_3) \subseteq X$,
- $\overline{Pre}_t(X) = \{s_2, s_3\}$ if $t \geq 1$ and $t \leq 2$ because $Post_t(s_2) \cap X \neq \emptyset$ and $Post_t(s_3) \cap X \neq \emptyset$.

It is easy to see that $\underline{Pre}_t(X)$ changes over time.

Finally, we give definition of the β-lower and β-upper predecessor anticipation for models in the form of timed transition systems.

Definition 5.6 Let $TTS(\mathscr{PM}) = (S, E, T, S_{init}, l, u)$ be a timed transition system modeling the plasmodium propagation in a given *Physarum* machine $\mathscr{PM}$, $X \subseteq S$ be a set of distinguished states, and $0 \leq \beta < 0.5$. The β-lower predecessor anticipation $\underline{Pre}_t^{\beta}(X)$ of X at the time instant t is defined as

$$\underline{Pre}_t^{\beta}(X) = \{s \in S : Post_t(s) \neq \emptyset \text{ and } Post_t(s) \overset{\beta}{\subseteq} X\}.$$

The β-upper predecessor anticipation $\overline{Pre}_t^{\beta}(X)$ of X at the time instant t is defined as

$$\overline{Pre}_t^{\beta}(X) = \{s \in S : 1 - \frac{Post_t(s) \cap X)}{Post_t(s)} < 1 - \beta\}.$$

Example 5.6 Let us consider a timed transition system $TTS(\mathscr{P}\mathscr{M}) = (S, E, T, S_{init}, l, u)$, from Example 3.7, modeling the plasmodium propagation in a given *Physarum* machine $\mathscr{P}\mathscr{M}$. Let us be interested in the following set of distinguished states $X = \{s_5, s_6, s_8\}$. For X and $\beta = 0.5$, we obtain:

- $\underline{Pre}_t^{0.5}(X) = \{s_2, s_3\}$ for each time instant t because $Post_t(s_2) \neq \emptyset$, $Post_t(s_2) \overset{0.5}{\subseteq} X$, $Post_t(s_3) \neq \emptyset$, and $Post_t(s_3) \overset{0.5}{\subseteq} X$ for each time instant t,
- $\overline{Pre}_t^{0.5}(X) = \{s_2, s_3\}$ for each time instant t.

Definition 5.7 Let $TTS(\mathscr{P}\mathscr{M}) = (S, E, T, S_{init}, l, u)$ be a timed transition system modeling the plasmodium propagation in a given *Physarum* machine $\mathscr{P}\mathscr{M}, X \subseteq S$, and $0 \leq \beta < 0.5$. The state s' is said to be:

- a continuous strict anticipator of the states from X if and only if

$$\underset{t \in \{t_0, t_1, t_2, \ldots\}}{\forall} s \in \underline{Pre}_t(X),$$

- an interim strict anticipator of the states from X if and only if s is not a continuous strict anticipator of the states from X, but

$$\underset{t \in \{t_0, t_1, t_2, \ldots\}}{\exists} s \in \underline{Pre}_t(X),$$

- a continuous quasi-strict anticipator of the states from X if and only if s is not a continuous strict anticipator of the states from X, but

$$\underset{t \in \{t_0, t_1, t_2, \ldots\}}{\forall} s \in \underline{Pre}_t^{\beta}(X),$$

- an interim quasi-strict anticipator of the states from X if and only if s is not a continuous and interim strict anticipator and continuous quasi-anticipator of the states from X, but

$$\underset{t \in \{t_0, t_1, t_2, \ldots\}}{\exists} s \in \underline{Pre}_t^{\beta}(X),$$

One can see that a continuous strict anticipator of the states from X always (i.e., at each time instant) anticipates the states from X whereas an interim strict anticipator of the states from X sometimes (not always) anticipates the states from X.

Example 5.7 Let us consider a timed transition system $TTS(\mathscr{P}\mathscr{M}) = (S, E, T, S_{init}, l, u)$, from Example 3.7, modeling the plasmodium propagation in a given *Physarum* machine $\mathscr{P}\mathscr{M}$. For $X = \{s_5, s_6, s_8\}$ and $\beta = 0.5$:

- s_3 is a continuous strict anticipator of the states from X because $s_3 \in \underline{Pre}_t(X)$ for each time instant t,
- s_2 is an interim strict anticipator of the states from X because $s_2 \in \underline{Pre}_t(X)$ only for $t \geq 1$ and $t \leq 2$ (i.e. s_2 is not a continuous strict anticipator),
- s_2 is a continuous quasi-strict anticipator of the states from X because $s_2 \in \underline{Pre}_t^{0.5}(X)$, but $s_2 \notin \underline{Pre}_t(X)$ for each time instant t.

5.3   Rough Set Descriptions Based on Complex Networks

In [40], rough set theory was used to analyze some properties of complex networks modeling eye-tracking sequences. In the analogous way, we can use rough set theory for analyzing behaviour of *Physarum* machines modeled by complex networks (see [58]).

Complex networks are networks whose structure is irregular, complex and dynamically evolving in time. Such properties can be observed in case of networks of protoplasmic veins formed by the plasmodium of *Physarum polycephalum*. Formally, a complex network can be presented as a graph either undirected or directed. In the proposed approach, we consider complex networks represented by undirected graphs. It means that we are not interested in directions of edges.

An undirected graph $G = (N, E)$ consists of two sets N and E such that $N \neq \emptyset$ and E is the set of unordered pairs of elements of N. The elements of $N = \{n_1, n_2, \ldots, n_q\}$ are the nodes of G, while the elements of $E = \{e_1, e_2, \ldots, e_r\}$ are the edges of G. The number of elements in N and E is denoted by q and r, respectively. The size of the graph is the number of nodes, i.e. q. In an undirected graph, each of the links is defined by a couple of nodes n_i and n_j, where $i, j = 1, \ldots, q$, and it is denoted as (n_i, n_j). The link is said to be incident on nodes n_i and n_j, or to join the two nodes. The two nodes joined by a link are referred to as adjacent or neighboring. For a graph G of size q, the number of edges r is at least 0 and at most $\frac{q(q-1)}{2}$ (when all the nodes are pairwise adjacent).

To build a model, in the form of a complex network $G = (N, E)$, of plasmodium propagation in a *Physarum* machine $\mathscr{PM} = \{Ph, Attr, Rep\}$, we can apply a procedure analogous to that proposed for a model in the form of a transition system (cf. Chap. 3). We take into consideration a stable state of $\mathscr{PM}$, i.e. the state at a given time instant t, when the set of all protoplasmic veins formed by plasmodium is fixed, i.e. $V = \{v_1, v_2, \ldots, v_{card(V)}\}$ (note that the superscript t has been omitted). The following objective functions are used:

- $\nu : Ph \cup Attr \rightarrow N$ assigning a node to each original point of plasmodium as well as to each attractant,
- $\epsilon : V \rightarrow E$ assigning an ege to each protoplasmic vein.

In the set of nodes of the complex network, we can distinguish some regions of interest (ROIs), i.e. selected subsets of nodes identified for a particular purpose. Let

$\Omega = \{\omega_1, \omega_2, \ldots, \omega_v\}$ be a set of all regions of interest identified in the complex network. $\mathcal{N} = \{N_{\omega_1}, N_{\omega_2}, \ldots, N_{\omega_v}\}$ denotes the family of sets of nodes corresponding to regions of interest.

For each node $n \in N_{\omega_1} \cup N_{\omega_2} \cup \cdots \cup N_{\omega_v}$, we define its inter-region neighborhood:

$$IRN(n) = \{n' : (n, n') \in E \wedge \underset{\omega \in \Omega}{\exists} (n' \in N_\omega \wedge n \notin N_\omega)\}.$$

On the basis of complex network modeling plasmodium propagation in a given *Physarum* machine $\mathcal{PM}$, we define a measure, derived from rough set theory, for assessing the cohesion of connections between regions of interest (cf. [40]).

Definition 5.8 Let $G = (N, E)$ be a complex network modeling the plasmodium propagation in a given *Physarum* machine $\mathcal{PM}$ and $\omega_i, \omega_j \subseteq N$ be two distinguished regions of interest. The lower approximation $\underline{IRN}(\omega_i \rightarrow \omega_j)$ of the inter-region neighborhood, from ω_i to ω_j, is defined as:

$$\underline{IRN}(\omega_i \rightarrow \omega_j) = \{n \in N_{\omega_i} : IRN(n) \neq \emptyset \wedge IRN(n) \subseteq N_{\omega_j}\}.$$

The upper approximation $\overline{IRN}(\omega_i \rightarrow \omega_j)$ of the inter-region neighborhood, from ω_i to ω_j, is defined as:

$$\overline{IRN}(\omega_i \rightarrow \omega_j) = \{n \in N_{\omega_i} : IRN(n) \cap N_{\omega_j} \neq \emptyset\}.$$

The lower approximation $\underline{IRN}(\omega_i \rightarrow \omega_j)$ of the inter-region neighborhood consists of all nodes N_{ω_i} which are connected by inter-region edges with nodes from N_{ω_j} only. The upper approximation $\overline{IRN}(\omega_i \rightarrow \omega_j)$ of the inter-region neighborhood consists of all nodes N_{ω_i} which are connected at least by one inter-region edge with nodes from N_{ω_j}.

The accuracy of approximation of the inter-region neighborhood can be defined analogously to the accuracy of approximation in rough set theory, i.e.:

$$\alpha_{IRN}(\omega_i \rightarrow \omega_j) = \frac{card\left(\underline{IRN}(\omega_i \rightarrow \omega_j)\right)}{card\left(\overline{IRN}(\omega_i \rightarrow \omega_j)\right)}.$$

We treat $\alpha_{IRN}(\omega_i \rightarrow \omega_j)$ as a measure of the cohesion of connection from the region of interest ω_i to the region of interest ω_j. If $\alpha_{IRN}(\omega_i \rightarrow \omega_j) = 1$, then the connection is the most coherent one.

It is worth noting that a measure of the cohesion of connections between regions of interest is not symmetrical, i.e. in general:

$$\alpha_{IRN}(\omega_i \rightarrow \omega_j) \neq \alpha_{IRN}(\omega_j \rightarrow \omega_i),$$

for $i, j = 1, 2, \ldots, v$ and $i \neq j$.

Fig. 5.1 A fragment of a complex network modeling plasmodium propagation in a given *Physarum* machine $\mathscr{PM}$

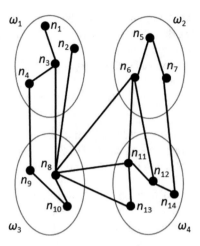

Example 5.8 Let us consider a fragment of a complex network, modeling the plasmodium propagation in a given *Physarum* machine $\mathscr{PM}$, shown in Fig. 5.1. The labels of edges have been omitted for better legibility. For instance, four regions of interest are identified with the following nodes:

- $N_{\omega_1} = \{n_1, n_2, n_3, n_4\}$,
- $N_{\omega_2} = \{n_5, n_6, n_7\}$,
- $N_{\omega_3} = \{n_8, n_9, n_{10}\}$,
- $N_{\omega_4} = \{n_{11}, n_{12}, n_{13}, n_{14}\}$.

A measure of the cohesion of connection from the region of interest ω_1 to the region of interest ω_3:

$$\alpha_{IRN}(\omega_1 \rightarrow \omega_3) = 1$$

because:

- $\underline{IRN}(\omega_1 \rightarrow \omega_3) = \{n_2, n_3, n_4\}$,
- $\overline{IRN}(\omega_1 \rightarrow \omega_3) = \{n_2, n_3, n_4\}$.

A measure of the cohesion of connection from the region of interest ω_3 to the region of interest ω_1:

$$\alpha_{IRN}(\omega_3 \rightarrow \omega_1) = \frac{1}{2}$$

because:

- $\underline{IRN}(\omega_3 \rightarrow \omega_1) = \{n_9\}$,
- $\overline{IRN}(\omega_1 \rightarrow \omega_3) = \{n_8, n_9\}$.

One can see that $\alpha_{IRN}(\omega_1 \rightarrow \omega_3) \neq \alpha_{IRN}(\omega_3 \rightarrow \omega_1)$.

We slightly modify definition of the lower and upper approximation of the interregion neighborhood in case of the Variable Precision Rough Set Model (VPRSM).

Definition 5.9 Let $G = (N, E)$ be a complex network modeling the plasmodium propagation in a given *Physarum* machine $\mathscr{PM}$, ω_i, $\omega_j \subseteq N$ be two distinguished regions of interest, and $0 \leq \beta < 0.5$. The β-lower approximation $\underline{IRN}(\omega_i \rightarrow \omega_j)$ of the inter-region neighborhood, from ω_i to ω_j, is given by:

$$\underline{IRN}^\beta(\omega_i \rightarrow \omega_j) = \{n \in N_{\omega_i} : IRN(n) \neq \emptyset \text{ and } IRN(n) \overset{\beta}{\subseteq} N_{\omega_j}\}.$$

The β-upper approximation $\overline{IRN}(\omega_i \rightarrow \omega_j)$ of the inter-region neighborhood, from ω_i to ω_j, is defined as:

$$\overline{IRN}^\beta(\omega_i \rightarrow \omega_j) = \{n \in N_{\omega_i} : 1 - \frac{card\,(IRN(n) \cap N_{\omega_j})}{card\,(IRN(n))} < 1 - \beta\}.$$

Now, one can see that the β-lower approximation of the inter-region neighborhood $\underline{IRN}(\omega_i \rightarrow \omega_j)$ consists of all nodes N_{ω_i} which are connected by inter-region edges, in most cases (i.e. in terms of the majority set inclusion), with nodes from N_{ω_j}.

A relaxed measure of the cohesion of connection from the region of interest ω_i to the region of interest ω_j has the form:

$$\alpha_{IRN}^\beta(\omega_i \rightarrow \omega_j) = \frac{card\,(\underline{IRN}^\beta(\omega_i \rightarrow \omega_j))}{card\,(\overline{IRN}^\beta(\omega_i \rightarrow \omega_j))}$$

for $0 \leq \beta < 0.5$.

Example 5.9 Let us consider a fragment of a complex network, modeling the plasmodium propagation in a given *Physarum* machine $\mathscr{PM}$, shown in Fig. 5.1 (see Example 5.8). A measure of the cohesion of connection from the region of interest ω_4 to the region of interest ω_2:

$$\alpha_{IRN}(\omega_4 \rightarrow \omega_2) = \frac{2}{3}$$

because:

- $\underline{IRN}(\omega_4 \rightarrow \omega_2) = \{n_{12}, n_{14}\}$,
- $\overline{IRN}(\omega_4 \rightarrow \omega_2) = \{n_{11}, n_{12}, n_{14}\}$.

Let $\beta = 0.5$, a relaxed measure of the cohesion of connection from the region of interest ω_4 to the region of interest ω_2:

$$\alpha_{IRN}^{0.5}(\omega_4 \rightarrow \omega_2) = 1$$

because:

- $\underline{IRN}^{0.5}(\omega_4 \rightarrow \omega_2) = \{n_{11}, n_{12}, n_{14}\}$,
- $\overline{IRN}^{0.5}(\omega_4 \rightarrow \omega_2) = \{n_{11}, n_{12}, n_{14}\}$.

5.4   Rough Set Descriptions Based on Tree Structures

In the specific case of plasmodium propagation in a given *Physarum* machine $\mathscr{PM}$, the graph structure (transition system, complex network, etc.) modeling the behaviour of $\mathscr{PM}$ can have the form of a tree. The root node (if exists) of the tree can correspond to the original point of plasmodium. In this section, we show how to deal with ambiguities in plasmodium motions using the rough set description of behaviour of a given *Physarum* machine $\mathscr{PM}$ built over the tree structure (see [59]). The proposed methodology is patterned upon the temporal logic of branching time [18]. In this logic, the underlying model is a tree of all possible computations. In case of *Physarum* machines, we can build a tree of all possible paths of plasmodium propagation.

A tree $\mathbf{T}$ is a partially ordered set (poset) $\mathbf{T} = (T, R_<)$ such that for each $x \in T$ the set $\{y : (y, x) \in R_<\}$ is well-ordered by the binary relation $R_<$.

For a given tree $\mathbf{T} = (T, R_<)$, we can consider its subtree $\mathbf{T}^x$ rooted at $x \in T$ denoted as $\mathbf{T}^x = (T^x, R_<)$ such that $T^x = \{y \in T : (x, y) \in R_< \text{ or } y = x\}$.

Let $\mathbf{T} = (T, R_<)$ be a tree. A segment $]a, b[$, where $a, b \in T$, is a set $]a, b[= \{x \in T : (a, x) \in R_< \text{ and } (x, b) \in R_<\}$. An element b is called a successor of an element a. An element a is called a predecessor of an element b. If $]a, b[= \emptyset$, then an element b is called an immediate successor of an element a and an element a is called an immediate predecessor of an element b.

Let $\mathbf{T} = (T, R_<)$ be a tree and $x \in T$. The set of all immediate successors of x is denoted by $Succ(x)$. The set of all immediate predecessors of x is denoted by $Pred(x)$. A leaf of $\mathbf{T}$ is any element $x \in T$ such that $Succ(x) = \emptyset$. A chain of $\mathbf{T}$ is any linearly ordered subset of T. A branch of $\mathbf{T}$ is any maximal (with respect to a number of elements) chain of $\mathbf{T}$. A set of all leaves of $\mathbf{T}$ is denoted by $Leaves(\mathbf{T})$. A set of all chains of $\mathbf{T}$ is denoted by $Chains(\mathbf{T})$. A set of all branches of $\mathbf{T}$ is denoted by $Branches(\mathbf{T})$.

To build a model, in the form of a tree $\mathbf{T} = (T, R_<)$, of behaviour of a given *Physarum* machine $\mathscr{PM} = \{Ph, Attr, Rep\}$, we can apply a procedure analogous to that proposed for a model in the form of a transition system (cf. Chap. 3). We take into consideration a stable state of $\mathscr{PM}$, i.e. the state at a given time instant t, when the set of all protoplasmic veins formed by plasmodium is fixed, i.e. $V = \{v_1, v_2, \ldots, v_{card(V)}\}$ (note that the superscript t has been omitted). It is necessary to underline that the elements of the tree (that is built) correspond only to active points in the *Physarum* machine. This remark is of importance for attractants. Only, attractants occupied by plasmodium are taken into consideration.

Let $\mathscr{PM} = \{Ph, Attr, Rep\}$ be the *Physarum* machine. The set of all attractants occupied by plasmodium at a given time instant t is denoted by $Attr_\bullet^t$. Obviously $Attr_\bullet^t \subseteq Attr$. The following objective functions are used:

- $\theta : Ph \cup Attr_\bullet^t \to T$ assigning an element of the tree $\mathbf{T}$ to each original point of plasmodium as well as to each attractant,
- $\rho : V \to R_<$ assigning an element of the relation $R_<$ to each protoplasmic vein.

Let $\mathbf{T} = (T, R_<)$ be a tree modeling behaviour of a given *Physarum* machine $\mathscr{PM}$ and $S \subseteq T$ be a set of distinguished elements of the tree $\mathbf{T}$. We can identify, in the set of elements of the tree $\mathbf{T}$, the following regions:

- $PosAnt^G(S)$ is a positive region of G-anticipation of elements from S.
- $BndAnt^G(S)$ is a boundary region of G-anticipation of elements from S.
- $NegAnt^G(S)$ is a negative region of G-anticipation of elements from S.
- $PosAnt^F(S)$ is a positive region of F-anticipation of elements from S.
- $BndAnt^F(S)$ is a boundary region of F-anticipation of elements from S.
- $NegAnt^F(S)$ is a negative region of F-anticipation of elements from S.
- $PosAnt^X(S)$ is a positive region of X-anticipation of elements from S.
- $BndAnt^X(S)$ is a boundary region of X-anticipation of elements from S.
- $NegAnt^X(S)$ is a negative region of X-anticipation of elements from S.

The division of regions given above corresponds to quantification over branches in the temporal logic of branching time.

Formal definitions of regions mentioned above are as follows.

Definition 5.10 Let $\mathbf{T} = (T, R_<)$ be a tree modeling the behaviour of a given *Physarum* machine $\mathscr{PM}$ and $S \subseteq T$. For each $x \in T$:

- $x \in PosAnt^G(S)$ if and only if

$$\forall_{B \in Branches(\mathbf{T}^x)} \forall_{y \in B} y \in S.$$

- $x \in BndAnt^G(S)$ if and only if

$$x \notin PosAnt^G(S) \text{ and } \exists_{B \in Branches(\mathbf{T}^x)} \forall_{y \in B} y \in S.$$

- $x \in NegAnt^G(S)$ if and only if

$$x \notin PosAnt^G(S) \text{ and } x \notin BndAnt^G(S).$$

- $x \in PosAnt^F(S)$ if and only if

$$\forall_{B \in Branches(\mathbf{T}^x)} \exists_{y \in B} y \in S.$$

- $x \in BndAnt^F(S)$ if and only if

$$x \notin PosAnt^F(S) \text{ and } \exists_{B \in Branches(\mathbf{T}^x)} \exists_{y \in B} y \in S.$$

- $x \in NegAnt^F(S)$ if and only if

$$x \notin PosAnt^F(S) \text{ and } x \notin BndAnt^F(S).$$

- $x \in PosAnt^X(S)$ if and only if

$$\underset{y \in Succ(x)}{\forall} \ y \in S.$$

- $x \in BndAnt^X(S)$ if and only if

$$x \notin PosAnt^X(S) \text{ and } \underset{y \in Succ(x)}{\exists} \ y \in S.$$

- $x \in NegAnt^X(S)$ if and only if

$$x \notin PosAnt^X(S) \text{ and } x \notin BndAnt^X(S).$$

One can see that:

- if $x \in Leaves(\mathbf{T})$ and $x \in S$, then $x \in PosAnt^G(S)$ and $x \in PosAnt^F(S)$,
- if $x \in Leaves(\mathbf{T})$ and $x \notin S$, then $x \in NegAnt^G(S)$ and $x \in NegAnt^F(S)$,
- if $x \in Leaves(\mathbf{T})$, then $x \in NegAnt^X(S)$,
- if $x \in PosAnt^G(S)$, then $x \in PosAnt^F(S)$,
- if $x \notin Leaves(\mathbf{T})$ and $x \in PosAnt^G(S)$, then $x \in PosAnt^F(S)$ and $x \in PosAnt^X(S)$.

Example 5.10 Let us consider a tree $\mathbf{T}(\mathscr{PM})$, modeling the plasmodium propagation in a given *Physarum* machine $\mathscr{PM}$, shown in Fig. 5.2. For the set

$$S = \{x_2, x_3, x_5, x_6, x_7, x_{11}, x_{12}, x_{13}\}$$

of distinguished elements of the tree $\mathbf{T}$, we obtain:

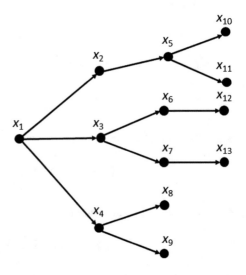

Fig. 5.2 A tree $\mathbf{T}(\mathscr{PM})$ modeling plasmodium propagation in a given *Physarum* machine $\mathscr{PM}$

- for x_1:

  - $x_1 \in NegAnt^G(S)$, because there is no branch in $Branches(\mathbf{T}^{x_1})$ satisfying a condition for G-anticipation,
  - $x_1 \in BndAnt^F(S)$, because there exist branches in $Branches(\mathbf{T}^{x_1})$, such as $\langle x_1, x_2, x_5, x_{10} \rangle$, $\langle x_1, x_2, x_5, x_{11} \rangle$, $\langle x_1, x_3, x_6, x_{12} \rangle$, and $\langle x_1, x_3, x_7, x_{13} \rangle$, satisfying a condition for F-anticipation, but there also exist branches in $Branches(\mathbf{T}^{x_1})$, such as $\langle x_1, x_4, x_8 \rangle$ and $\langle x_1, x_4, x_9 \rangle$, not satisfying this condition,
  - $x_1 \in BndAnt^X(S)$, because $Succ(x_1) = \{x_2, x_3, x_4\}$ and $x_2, x_3 \in S$, but $x_4 \notin S$,

- for x_2:

  - $x_2 \in BndAnt^G(S)$, because $\langle x_2, x_5, x_{11} \rangle \in Branches(\mathbf{T}^{x_2})$ satisfies a condition for G-anticipation, but $\langle x_2, x_5, x_{10} \rangle \in Branches(\mathbf{T}^{x_2})$ does not satisfy this condition,
  - $x_2 \in PosAnt^F(S)$, because all branches in $Branches(\mathbf{T}^{x_2})$, i.e. $\langle x_2, x_5, x_{10} \rangle$ and $\langle x_2, x_5, x_{11} \rangle$, satisfy a condition for F-anticipation,
  - $x_2 \in PosAnt^X(S)$, because $Succ(x_2) = \{x_5\}$ and $x_5 \in S$,

- for x_3:

  - $x_3 \in PosAnt^G(S)$, because all branches in $Branches(\mathbf{T}^{x_3})$, i.e. $\langle x_3, x_6, x_{12} \rangle$ and $\langle x_3, x_7, x_{13} \rangle$, satisfy a condition for G-anticipation,
  - $x_3 \in PosAnt^F(S)$, because $x_3 \in PosAnt^G(S)$,
  - $x_3 \in PosAnt^X(S)$, because $x_3 \notin Leaves(\mathbf{T})$ and $x_3 \in PosAnt^G(S)$,

- for x_4:

  - $x_4 \in NegAnt^G(S)$, because there is no branch in $Branches(\mathbf{T}^{x_4})$ satisfying a condition for G-anticipation,
  - $x_4 \in NegAnt^F(S)$, because there is no branch in $Branches(\mathbf{T}^{x_4})$ satisfying a condition for F-anticipation,
  - $x_4 \in NegAnt^X(S)$, because $Succ(x_4) = \{x_8, x_9\}$, but $x_8, x_9 \notin S$,

- for x_5:

  - $x_5 \in BndAnt^G(S)$, because $\langle x_5, x_{11} \rangle \in Branches(\mathbf{T}^{x_5})$ satisfies a condition for G-anticipation, but $\langle x_5, x_{10} \rangle \in Branches(\mathbf{T}^{x_5})$ does not satisfy this condition,
  - $x_5 \in PosAnt^F(S)$, because all branches in $Branches(\mathbf{T}^{x_5})$, i.e. $\langle x_5, x_{10} \rangle$ and $\langle x_5, x_{11} \rangle$, satisfy a condition for F-anticipation,
  - $x_5 \in BndAnt^X(S)$, because $Succ(x_5) = \{x_{10}, x_{11}\}$ and $x_{11} \in S$, but $x_{10} \notin S$,

- for x_6:

  - $x_6 \in PosAnt^G(S)$, because all branches in $Branches(\mathbf{T}^{x_6})$, i.e. $\langle x_6, x_{12} \rangle$, satisfy a condition for G-anticipation,
  - $x_6 \in PosAnt^F(S)$, because $x_6 \in PosAnt^G(S)$,
  - $x_6 \in PosAnt^X(S)$, because $x_6 \notin Leaves(\mathbf{T})$ and $x_6 \in PosAnt^G(S)$,

- for x_7:
  - $x_7 \in PosAnt^G(S)$, because all branches in $Branches(\mathbf{T}^{x_7})$, i.e. $\langle x_7, x_{13} \rangle$, satisfy a condition for G-anticipation,
  - $x_7 \in PosAnt^F(S)$, because $x_7 \in PosAnt^G(S)$,
  - $x_7 \in PosAnt^X(S)$, because $x_7 \notin Leaves(\mathbf{T})$ and $x_7 \in PosAnt^G(S)$,

- for x_8:
  - $x_8 \in NegAnt^G(S)$, because $x_8 \in Leaves(\mathbf{T})$ and $x_8 \notin S$,
  - $x_8 \in NegAnt^F(S)$, because $x_8 \in Leaves(\mathbf{T})$ and $x_8 \notin S$,
  - $x_8 \in NegAnt^X(S)$, because $x_8 \in Leaves(\mathbf{T})$,

- for x_9:
  - $x_9 \in NegAnt^G(S)$, because $x_9 \in Leaves(\mathbf{T})$ and $x_9 \notin S$,
  - $x_9 \in NegAnt^F(S)$, because $x_9 \in Leaves(\mathbf{T})$ and $x_9 \notin S$,
  - $x_9 \in NegAnt^X(S)$, because $x_9 \in Leaves(\mathbf{T})$,

- for x_{10}:
  - $x_{10} \in NegAnt^G(S)$, because $x_{10} \in Leaves(\mathbf{T})$ and $x_{10} \notin S$,
  - $x_{10} \in NegAnt^F(S)$, because $x_{10} \in Leaves(\mathbf{T})$ and $x_{10} \notin S$,
  - $x_{10} \in NegAnt^X(S)$, because $x_{10} \in Leaves(\mathbf{T})$,

- for x_{11}:
  - $x_{11} \in PosAnt^G(S)$, because $x_{11} \in Leaves(\mathbf{T})$ and $x_{11} \in S$,
  - $x_{11} \in PosAnt^F(S)$, because $x_{11} \in Leaves(\mathbf{T})$ and $x_{11} \in S$,
  - $x_{11} \in NegAnt^X(S)$, because $x_{11} \in Leaves(\mathbf{T})$,

- for x_{12}:
  - $x_{12} \in PosAnt^G(S)$, because $x_{12} \in Leaves(\mathbf{T})$ and $x_{12} \in S$,
  - $x_{12} \in PosAnt^F(S)$, because $x_{12} \in Leaves(\mathbf{T})$ and $x_{12} \in S$,
  - $x_{12} \in NegAnt^X(S)$, because $x_{12} \in Leaves(\mathbf{T})$,

- for x_{13}:
  - $x_{13} \in PosAnt^G(S)$, because $x_{13} \in Leaves(\mathbf{T})$ and $x_{13} \in S$,
  - $x_{13} \in PosAnt^F(S)$, because $x_{13} \in Leaves(\mathbf{T})$ and $x_{13} \in S$,
  - $x_{13} \in NegAnt^X(S)$, because $x_{13} \in Leaves(\mathbf{T})$.

Thus, rough sets allow us to formalize the slime mould motions with propagation in all possible directions. Meanwhile, we can consider rough sets on transitions or on trees. In any case, as we see, the reality appears for the slime mould as something changeable, instable or even ambiguous. Such reality (universe) is called non-well-founded in the terms of mathematics.

Chapter 6
Non-Well-Foundedness

6.1 Non-Well-Founded Reality

A non-well-founded set theory belongs to axiomatic set theories that violate the rule of well-foundedness and, as an example, allow sets to contain themselves: $X \in X$. In non-well-founded set theories [3], the foundation axiom of Zermelo-Fraenkel set theory is replaced by axioms implying its negation. The theory of non-well-founded sets has been explicitly applied in diverse fields such as logical modelling non-terminating computational processes and behaviour of interactive systems in computer science (process algebra, coalgebra, logical programming based on coinduction and corecursion), linguistics and natural language semantics (situation theory), logic (analysis of semantic paradoxes). So, if we are going to formalize the plasmodium propagation, we should define a non-well-founded universe for modelling the slime mould transitions.

Non-well-founded sets have been implicitly used in non-standard (more precisely, *non-Archimedean*) analysis like infinitesimal [70] and *p*-adic analysis [16, 50, 52]. The point is that denying the foundation axiom in number systems implies setting the *non-Archimedean ordering structure* [46]. Recall that Archimedes' axiom affirms the existence of an integer multiple of the smaller of two numbers which exceeds the greater: for any positive real or rational number y, there exists a positive integer n such that $y > 1/n$ or $ny > 1$. The informal sense of Archimedes' axiom is that anything can be measured by a rigid scale. Refusing the Archimedean axiom entails the existence of infinitely large numbers (in the case of field this means additionally the existence of infinitely small numbers). So, there is the following true implication: if the field has a Dedekind completeness, which affirms the existence of a supremum for every bounded set, then this field has the Archimedean property. It is known that the field $\mathbf{R}$ of real numbers satisfies the Dedekind completeness (i.e. any nonempty set of real numbers which has an upper bound has a least upper bound), then it satisfies Archimedes' axiom, too. Refusing the latter implies refusing the standard completeness. Meanwhile, there is no upper bound for the set $\mathbf{Z} \subset \mathbf{R}$ of all integers;

© Springer International Publishing AG, part of Springer Nature 2019
A. Schumann and K. Pancerz, *High-Level Models of Unconventional Computations*, Studies in Systems, Decision and Control 159,
https://doi.org/10.1007/978-3-319-91773-3_6

no matter how large a number we choose for the upper bound, there will always be some integer bigger than it. Without the standard completeness we obtain infinite real numbers (infinitely large real numbers) and infinitesimals (infinitely small real numbers). For them there are no least upper bounds in the general case. The field with infinite numbers and infinitesimals is called the field of hyperreal numbers and denoted by $^*\mathbf{R}$. This field is not complete in the standard sense, because, for example, the set of infinitesimals does not have a least upper bound. On the other hand, the set $\mathbf{N}$ of positive integers is bounded above by the member $t \in {}^*\mathbf{N}\backslash\mathbf{N}$, where $^*\mathbf{N}$ is the set of hypernatural numbers and t is the sequence given by $t_n = n$ for all $n \in \mathbf{N}$. But $\mathbf{N}$ can have no least upper bound: if $n \le c$ for all $n \in \mathbf{N}$ then $n \le c - 1$ for all $n \in \mathbf{N}$.

There exists also a different version of mathematical analysis in that Archimedes' axiom is rejected, namely, p-adic analysis (e.g., see [50]). In this analysis, one investigates the properties of the completion of the field $\mathbf{Q}$ of rational numbers (respectively, the properties of the completion of the ring $\mathbf{Z}$ of integers) with respect to the metric $\rho_p(x, y) = |x - y|_p$, where the norm $| \cdot |_p$ called p-adic is defined as follows: (1) $|y|_p = 0 \leftrightarrow y = 0$, (2) $|x \cdot y|_p = |x|_p \cdot |y|_p$, (3) $|x + y|_p \leqslant \max(|x|_p, |y|_p)$ (non-Archimedean triangular inequality). That metric over the field $\mathbf{Q}$ (respectively, over the ring $\mathbf{Z}$) is non-Archimedean, because $|n \cdot 1|_p \leqslant 1$ for all $n \in \mathbf{Z}$. This completion of the field $\mathbf{Q}$ (respectively, of the ring $\mathbf{Z}$) is called the field $\mathbf{Q}_p$ of p-adic numbers (respectively, the ring $\mathbf{Z}_p$ of p-adic integers). $\mathbf{Q}_p$ is not an ordered field. This means that there is no total ordering of its elements that agrees with the field operations. On the other hand, the ring $\mathbf{Z}_p$ includes $\mathbf{Z}$ and satisfies the equality $\mathbf{Z}_p = \{x \in \mathbf{Q}_p : |x \cdot 1|_p \leqslant 1\}$. As a result, in $\mathbf{Z}_p$ there are integers that are naturally to be interpreted as infinitely large numbers.

A connection between denying the axiom of foundation and denying the Archimedean axiom may be shown as follows. Recall that a binary relation, R, is well-founded on X if and only if every non-empty subset of X has a minimal element with respect to R. In other words, if R is well-founded, then the *induction principle* is valid: if R is a well-founded relation on X and $P(x)$ is some property of elements of X, then to show $P(x)$ holds for all elements of X, it suffices to show that if x is a member of X and $P(y)$ is true for all y such that yRx, then $P(x)$ must also be true: $\forall x \in X((\forall y \in X(yRx \to P(y))) \to P(x)) \to \forall x \in XP(x)$. For instance, if a membership relation, $\in$, is well-founded, then the epsilon-induction ($\in$-induction) holds, i.e. for any formula φ:

$$\forall x((\forall y(y \in x \to \varphi(y))) \to \varphi(x)) \to \forall x\varphi(x).$$

More generally, one can define objects by induction on any well-founded relation R. This generalized kind of induction is called sometimes Noetherian induction.

The negation of the axiom of foundation causes that there are objects that cannot be defined by Noetherian induction. This means that there exists a non-empty subset of X that has no minimal element with respect to a relation R. The latter is called a non-well-founded relation. We remember that there are no least upper bounds for non-Archimedean numbers in the general case. This means that those sets of upper bounds have no least (minimal) element. Therefore we cannot use Noetherian induction for the set of infinitesimals or for the set of infinitely large integers (notice that sets $^*\mathbf{Z} \backslash \mathbf{Z}$ and $\mathbf{Z}_p \backslash \mathbf{Z}$ of infinitely large integers have a different meaning). For example, there are no $\in$-induction for these sets. Taking into account this circumstance, we can state that some initial objects of non-Archimedean mathematics are objects obtained implicitly by denying the axiom of foundation. Non-Archimedean numbers are non-well-founded.

Instead of Noetherian induction, we can use coinduction as the dual notion applied to non-well-founded objects. In this case, a binary relation, R, is non-well-founded on X if and only if every non-empty subset of X has a maximal element with respect to R (this definition is one of the possible formulation of anti-foundation axiom). Transfinite coinduction is implicitly used in the definition of spherically complete ultrametric spaces: the ultrametric field K is spherically complete if and only if it is maximal among ultrametric fields with the same value group and residue field.

The conventional approach to modeling statistical phenomena uses classical (Kolmogorov's) probability theory built in the language of well-founded mathematics of real numbers. It sets a framework of modern physics, taking into account that physical reality is regarded in modern science as reality of stable repetitive phenomena (phenomena that have probabilities, i.e. do not fluctuate in the standard real metric). However, we can assume another approach to the physical measurement for statistical phenomena that uses tools of non-well-founded mathematics, in particular uses the coinduction principle and non-well-founded probability theory on non-Archimedean structures.

6.2 Algorithms Versus Coalgorithms, Induction Versus Coinduction

Beyond all doubt, the most basic notion of mathematics and physics is an algorithm. It plays a significant role providing, e.g. a correct (from the standpoint of logic) reasoning in mathematics and a well-defined measurement by rigid scales in physics. Its simplest definition is as follows: the algorithm is a set of instructions for solving a problem. In computer sciences, the algorithm is regarded either as the one implemented by a computer program or simulated by a computer program. In other words, the algorithm is reduced to the computer's process instructions, telling the computer what specific steps and in what specific order to perform in order to carry out a specified task.

There are at least two general approaches to explicate mathematically the notion of algorithm: conventional and unconventional. The *conventional* approach is presented by the following well-known statement: every algorithm can be simulated by a Turing machine and formally expressed by the lambda calculus. This statement is known as the *Church-Turing thesis*. Recall that it was initially formulated as follows: the intuitive notion of effective computability for functions and algorithms is well explicated by Turing machines (Turing) or the lambda calculus (Church). This thesis equated lambda calculus, Turing machines and algorithmic computing as equivalent mechanisms of problem solving. Later, in the conventional approach to algorithms, it was reinterpreted as a uniform complete mechanism for solving all computational problems.

Notice that the standard name of Turing machines actually refers, in Turing's words, to automatic machines, or a-machines. He also proposed other models of computation: c-machines (choice machines) and u-machines (unorganized machines). Turing argued for the claim (Turing's thesis) that whenever there is an effective method for obtaining the values of a mathematical function, the function can be computed by a Turing a-machine. At the same time, Church formulated the following thesis: a function of positive integers is effectively calculable only if it is recursive. If attention is restricted to functions of positive integers then Church's thesis and Turing's thesis are equivalent. It is important to distinguish between the Turing-Church thesis and the different proposition that whatever can be calculated by a machine can be calculated by a Turing machine [27]. The two propositions are sometimes confused. Gandy termed the second proposition *'Thesis M'*: whatever can be calculated by a machine is Turing-machine-computable [36]. The latter thesis is a fundamental of conventional approach for explicating the notion of algorithm.

So, conventionally, *the algorithm is associated with step-by-step processing information and with mapping the initial input to the final output, ignoring the external world while it executes.* As we see, we accept here the five of requirements for an algorithm:

1. the finiteness, an algorithm must always terminate after a finite number of steps;
2. the definiteness, each step of an algorithm must be precisely defined;
3. the determinateness by input (initial data), i.e. by quantities which are given to it initially before the algorithm begins;
4. the finiteness of output (desired result), i.e. of quantities which have a specified relation to the inputs, they have no history dependence for multiple computations;
5. the effectiveness, all of the operations to be performed in the algorithm can in principle be done exactly and in a finite length of time.

Recall that Markov formulated the following three features characteristic of algorithms [53]:

1. the definiteness of the algorithm, consisting in the universal comprehensibility and precision of prescription, leaving no place to arbitrariness;
2. the generality of the algorithm, the possibility of starting out with initial data, which may vary within given limits;
3. the conclusiveness of the algorithm, the orientation of the algorithm toward obtaining some desired result, which is indeed obtained in the end with proper initial data.

The most basic notion in the conventional approach to algorithms implicitly used here is the least fixed point presupposing the *induction principle* (Noetherian induction), e.g. this notion is fundamental for *recursion*.

However, this notion is violated fully in the *unconventional approach* to explicate the notion of algorithm. The conventional treatment of algorithm is unapplied, broken in unconventional computing models such as computing on the medium of *Physarum polycephalum*. The main originality of unconventional computing is that just as conventional models of computation make a distinction between the structural part of a computer, which is fixed, and the data on which the computer operates, which are variable, so unconventional models assume that both structural parts and computing data are variable ones [4]. Therefore the essential point of conventional computation is that the physics is segregated once and for all within the logic primitives. For instance, reaction-diffusion computing [7] is one of the unconventional models built on spatially extended chemical systems, which process information using interacting growing patterns, of excitable and diffusive waves. In reaction-diffusion processors, both the data and the results of the computation are encoded as concentration profiles of the reagents. The computation is performed via the spreading and interaction of wave fronts.

Unconventional computing may be presented as massive-parallel locally-connected mathematical machines. They cannot have a single centralized source exercising precise control over vast numbers of heterogeneous devices. The *Physarum* machines are decentralized and massive-parallel, too. Therefore the conventional definition of algorithm for *Physarum* motions is unapplied. Instead of recursion, we can apply there corecursion as a type of operation that is dual to recursion [73, 74]. Corecursion is typically used to generate infinite data structures. The rule for primitive corecursion on codata is the dual to that for primitive recursion on data. Instead of descending on the argument, we ascend on the result. Notice that corecursion creates potentially infinite codata, whereas ordinary recursion analyzes necessarily finite data. By induction and recursion we use the notion of least fixed points, whereas by coinduction and corecursion we use the notion of greatest fixed points.

Induction and recursion are firmly entrenched as fundamentals for proving properties of inductively defined objects, e.g. of finite or enumerable objects. Discrete mathematics and computer science abound with such objects, and mathematical induction is certainly one of the most important tools. However, we cannot use the principle of induction for non-well-founded objects. Instead of this principle, the notion of *coinduction* appears as the dual to induction.

The difference between induction and coinduction may be well defined as follows. Firstly, let an operation $\Phi : \mathscr{P}(A) \to \mathscr{P}(A)$, where $\mathscr{P}(A)$ is the powerset of A, be defined as monotone iff $X \subseteq Y$ implies $\Phi(X) \subseteq \Phi(Y)$ for $X, Y \subseteq A$. Any monotone operation Φ has the least and the greatest fixed point, X_Φ and X^Φ respectively, that is, $\Phi(X_\Phi) = X_\Phi$, $\Phi(X^\Phi) = X^\Phi$, and for any other fixed point $Y \subseteq A$ of Φ (i.e. $\Phi(Y) = Y$) we have $X_\Phi \subseteq Y \subseteq X^\Phi$. The sets X_Φ and X^Φ can be defined by $X_\Phi :: = \bigcap \{Y : Y \subseteq A, \Phi(Y) \subseteq Y\}$, $X^\Phi :: = \bigcup \{Y : Y \subseteq A, Y \subseteq \Phi(Y)\}$. It is easy to see that the monotonicity of Φ implies the required properties of X_Φ and X^Φ.

On the one hand, by definition of X_Φ, we have for any set $Y \subseteq A$ that $\Phi(Y) \subseteq Y$ implies $X_\Phi \subseteq Y$. This principle is called *induction*. On the other hand, by definition of X^Φ, we have for any set $Y \subseteq A$ that $Y \subseteq \Phi(Y)$ implies $Y \subseteq X^\Phi$. This principle is called *coinduction*.

Thus, the difference of applying two kinds of fixed points may by illustrated as follows. The least fixed point of the equation $S = A \times S$ is the empty set (i.e. it is so from the standpoint of conventional approach), while the greatest fixed point is the set of all streams over S (i.e. it is so from the standpoint of unconventional approach).

The greatest fixed point (respectively, coinduction and corecursion) allows us to describe the behaviour of computing without a priori presuppositions. As a result, unconventionally, *the algorithm can be viewed as processing information that may be massive-parallel and as mapping the initial input to the final output, whose values depend on interaction with the open unpredictable environment*, i.e. identical inputs may provide different outputs, as the system learns and adapts to its history of inter-actions. The algorithm with such interpretation is said to be *coalgorithm* (it is an algorithm based on coinduction and corecursion). Its basic properties are as follows:

1. the infiniteness, an algorithm is simulated behaviourally by differential equations;
2. the indefiniteness, there can be a massive-parallel change of system;
3. the determinateness not only by input (initial data), but also by interaction with the open unpredictable environment;
4. the infiniteness of output;
5. the bisimulation[1] effectiveness, when there is no prespecified endpoint.

The notion of coalgorithm appeared due to applications of non-well-founded mathematics in computer sciences proposed within the framework of *interactive-computing* or *concurrency paradigm*, e.g. the latter assumes coinductive and core-cursive methods to be used in computations. Although in such a paradigm, compu-tation models may be abstract in the same measure as Turing machines, they assume a combination of physics and logic in the computation, as well as that in cellular automata or in other unconventional computing models. In conventional computing, the computation is performed in a closed-box fashion, transforming a finite input, determined by the start of the computation, to a finite output, available at the end of the computation, in a finite amount of time. Therefore physical implementation does not play role in such computation that may be considered as process in a black box. As opposed to the conventional approach, in interactive models inputs and outputs may be infinite and computation in each point of lattice may proceed simultaneously and independently.

Massive-parallel locally-connected mathematical machines, such as *Physarum* machines, cannot have a single centralized source. Interactive-computing paradigm is able to describe concurrent or parallel computations whose configuration may change during the computation and is decentralized as well. Within the framework of this paradigm, so-called concurrency calculi also called process algebras are built. They are typically presented using systems of equations. These formalisms for concurrent systems are formal in the sense that they represent systems by expressions and then reason about systems by manipulating the corresponding expressions.

---

[1]Bisimulation is a binary relation between labelled transition systems with the same set of actions, associating those systems if one of them simulates the other and vice-versa. According to bisimu-lation, two systems are equivalent if both have the same behaviour.

One of the most useful non-well-founded mathematical objects defined by coinduction is a stream—a corecursive data-type of the form $s = (a, s')$, where s' is another stream. A stream can be exemplified as a succession of days, unfolding in a cyclic pattern:

$$days = (Monday, (Tuesday, (Wednesday, \ldots (Sunday, days) \ldots))).$$

It is an example of the explicit non-well-founded object that has a self-reference structure. 'Days' is not a finite set {Monday, ..., Sunday}, because it has an infinite number of members of the kind 'Monday', of the kind 'Sunday', etc. Therefore it is natural to represent 'days' as an infinite tuple (Monday, (..., (Sunday, (Monday, (...))))). The stream of such a kind is circular: after 'Sunday' we repeat again and again: 'Monday', ..., 'Sunday'. A circular stream we cannot define by means of conventional mathematics. For example, in conventional mathematics there is the Russellian paradox that appears if we have $a \in a$, here we obtain a circular definition too: $a = \{a\}$, then $a = \{a\} = \{\{a\}\} = \{\{\{a\}\}\} = \ldots$

Let us consider some opportunities of using the greatest-fixed point notion in the statistical approach to measurement [45, 48]. To do it, let us make some basic definitions of probability applications of the greatest-fixed point measurement. In the sequel we will use coalgorithms, more precisely the coinductive principle.

Let A be any set. We define the set A^ω of all streams over A as $A^\omega = \{\sigma : \{0, 1, 2, \ldots\} \to A\}$. For a stream σ, we call $\sigma(0)$ the initial value of σ. We define the *derivative* of a stream σ, for all $n \geq 0$, by $\sigma'(n) = \sigma(n + 1)$. For any $n \geq 0$, $\sigma(n)$ is called the n-th element of σ. It can also be expressed in terms of higher-order stream derivatives, defined, for all $k \geq 0$, by $\sigma^{(0)} = \sigma$; $\sigma^{(k+1)} = (\sigma^{(k)})'$. In this case the n-th element of a stream σ is given by $\sigma(n) = \sigma^{(n)}(0)$. Also, the stream is understood as an infinite sequence of derivatives. It will be denoted by an infinite sequence of values or by an infinite tuple: $\sigma = \sigma(0) :: \sigma(1) :: \sigma(2) :: \cdots :: \sigma(n - 1) :: \sigma^{(n)}$, $\sigma = (\sigma(0), \sigma(1), \sigma(2), \ldots)$. The state stream $a::a::a:: \ldots$ is denoted by $[a]$.

It can be easily shown that p-adic numbers may be represented as potentially infinite data structures such as streams. Each stream of the form $\sigma = \sigma(0)::\sigma(1)::\sigma(2):: \ldots :: \sigma(n - 1)::\sigma^{(n)}$, where $\sigma(n) \in \{0, 1, \ldots, p - 1\}$ for every $n \in \mathbf{N}$, may be converted into a p-adic integer by the following rule:

$$\forall n \in \mathbf{N}, \sigma(n) = \sum_{k=0}^{n} \sigma(k) \cdot p^k \wedge \sigma(n) = \sigma(0) :: \sigma(1) :: \cdots :: \sigma(n). \qquad (6.1)$$

And vice versa, each p-dic integer may be converted into a stream taking rule (6.1). Such a stream is called p-adic.

Streams are defined by coinduction: two streams σ and τ in A^ω are equal if they are *bisimilar*: (i) $\sigma(0) = \tau(0)$ (they have the same *initial value*) and (ii) $\sigma' = \tau'$ (they have the same *differential equation*). To set addition and multiplication by coinduction, we should use the following facts about differentiation of sums and

Table 6.1 Coinductive definitions of sum, product and inverse

Differential equation	Initial value	Name
$(\sigma + \tau)' = \sigma' + \tau'$	$(\sigma + \tau)(0) = \sigma(0) + \tau(0)$	Sum
$(\sigma \times \tau)' =$ $(\lvert\sigma(0)\rvert \times \tau') + (\sigma' \times \tau)$	$(\sigma \times \tau)(0) = \sigma(0) \times \tau(0)$	Product
$(\sigma^{-1})' =$ $\lvert -1 \rvert \times \lvert\sigma(0)\rvert^{-1} \times \sigma' \times \sigma^{-1}$	$(\sigma^{-1})(0) = \sigma(0)^{-1}$	Inverse

products by applying the basic operations: $(\sigma + \tau)' = \sigma' + \tau', (\sigma \times \tau)' = (\lvert\sigma(0)\rvert \times \tau') + (\sigma' \times \tau)$, where $\lvert\sigma(0)\rvert = \langle\sigma(0), 0, 0, 0, \dots\rangle$. Now we can define them as well as another stream operation, see Table 6.1.

So, we examine streams as dynamical entities, whose behaviour consists of the repeatedly offering of the next element of the stream. Using coinduction streams are defined by specifying their behaviour, and the equality of two streams can be established by proving that they have the same behaviour (in other words, that they are 'behaviourally' equivalent).

The main reason for choosing to use streams as the coinductive datatype consists in a possibility to give definitions very close in style to those with self-reference corresponding to $\sigma = a{::}\sigma$.

We can try to get non-well-founded probabilities on a non-well-founded algebra $F^V(A^\omega)$ of fuzzy subsets $Y \subset A^\omega$ that consists of the following [83]: (1) union, intersection, and difference of two *non-well-founded fuzzy* subsets of A^ω; (2) $\emptyset$ and A^ω. In this case a *finitely additive non-well-founded probability measure* is a nonnegative set function $P(\cdot)$ defined for sets $Y \in F^V(A^\omega)$ that runs the set V (for example, $V = \mathbf{Z}_p$) and satisfies the following properties: (1) $P(A) \geq [0]$ for all $A \in F^V(A^\omega)$, (2) $P(A^\omega) = \lvert 1 \rvert$ and $P(\emptyset) = [0]$, (3) if $A \in F^V(A^\omega)$ and $B \in F^V(A^\omega)$ are disjoint, then $P(A \cup B) = P(A) + P(B)$, (4) $P(\neg A) = \lvert 1 \rvert + \lvert -1 \rvert \times P(A)$ for all $A \in F^V(A^\omega)$.

This probability measure is called *non-well-founded probability*. Their main originality is that conditions 3, 4 are independent. As a result, in a probability space $(X, F^V(X), P)$ some Bayes' formulas do not hold in the general case.

Suppose that the ordering relation on $\mathbf{Z}_p$ is defined digit by digit. Then the number -1 is the greatest in $\mathbf{Z}_p$. As an example of *trivial non-well-founded probability* we can introduce the following function defined on p-adic streams by coinduction: $P(\sigma) = \inf(\sigma, [-1]) \times [-1]^{-1}$ for every $\sigma \in \mathbf{Z}_p$. Notice that p-adic probabilities used in adelic or p-adic quantum mechanics are particular cases of non-well-founded probabilities.

If we take the p-adic case of non-well-founded probability theory, then we observe essentially new properties of relative frequencies that do not appear on real numbers. For example, consider two attributes α_1 and α_2. Suppose that in the first $N{::} = N_k = (\sum_{j=0}^{k} 2^j)^2$ tests the label α_1 has $n_N(\alpha_1; x) = 2^k$ realizations,

α_2 has $n_N(\alpha_2; x) = \sum_{j=0}^{k} 2^j$ realizations. According to our intuition, their probabilities should be different, but in real probability theory we obtain: $P_x(\alpha_1) = \lim n_N(\alpha_1; x)/N = P_x(\alpha_2) = \lim n_N(\alpha_2; x)/N = 0$. In 2-adic probability theory we have $P_x(\alpha_1) = 0 \neq P_x(\alpha_2) = -1$, because in $\mathbf{Q}_2$, $2^k \to 0$, $k \to 0$, and $-1 = 1 + 2 + 2^2 + \cdots + 2^n + \cdots$.

This example shows that in p-adic probability theory there are statistical phenomena for that relative sequences of observed events have non-zero probabilities in the p-adic metric, but do not have positive probabilities in the standard real metric.

We assume that reality is non-well-founded and the *Physarum* motions can be formalized as some streams defined coinductivelly. Therefore, the slime mould motions can be programmed only within an object-oriented approach that we are going define right now.

Chapter 7
Physarum Language

7.1 Object-Oriented Programming Language

To build high-level models of plasmodium propagation in *Physarum* machines and analyze their behaviours in the non-well-founded universe, an object-oriented programming language, called the *Physarum* language, has been developed by us (see [62]). The *Physarum* language can be classified to the category of *prototype-based languages*. The prototype-based approach [28] is also called class-less or instance-based programming because prototype-based languages are based upon the idea that objects that represent individuals can be created without reference to class-defining. The objects that are manipulated at runtime are the prototypes. In the *Physarum* language, there are, among others, inbuilt sets of prototypes corresponding to high-level models used for describing the plasmodium propagation in *Physarum* machines (e.g. ladder diagrams, transition systems, timed transition systems, Petri nets, tree structures). Objects can be instantiated via the keyword *new* using defined constructors. Methods are used to manipulate features of the objects and create relationships between objects.

Each high-level model can be translated into the low-level language, i.e. a spatial distribution of stimuli (attractants and/or repellents). Such distribution can be treated as a program for the *Physarum* machine. Some examples of geometrical distributions of stimuli were shown in Chap. 4. For the programming purpose, a compiler embodied in a computer tool called *PhysarumSoft* (see Chap. 12) translates the high-level models built in the *Physarum* language into the spatial distribution (configuration) of stimuli (attractants and/or repellents) for *Physarum* machines. It is worth noting, that, as it is shown in Fig. 7.1, the high-level models built in the *Physarum* language are also used to determine some ambiguities in plasmodium motions in *Physarum* machines using rough set based descriptions (see Chap. 5).

In the next sections, the *Physarum* language for three high-level models, namely Petri nets, transition systems, and tree structures, is outlined.

© Springer International Publishing AG, part of Springer Nature 2019
A. Schumann and K. Pancerz, *High-Level Models of Unconventional Computations*, Studies in Systems, Decision and Control 159,
https://doi.org/10.1007/978-3-319-91773-3_7

Fig. 7.1 Appliacations of the *Physarum* language

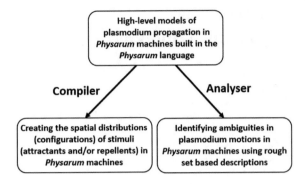

7.2 Physarum Language for Petri Net Models

In case of building Petri net models, two modes are available in the *Physarum* language:

- place-transition-arc mode,
- transition input/output mode.

The main prototypes defined in the *Physarum* language and their selected methods are collected in Tables 7.1 and 7.2 for the place-transition-arc mode and transition input/output mode, respectively.

Example 7.1 Let us consider the Petri net model of the NOT gate described in Sect. 4.2. We can define a Petri net structure $PNS_{not} = (Pl_{not}, Tr_{not}, Arc_{not}, w_{not})$ for the NOT gate:

- $Pl_{not} = \{P, R_x, A_y\}$,
- $Tr_{not} = \{T\}$,
- $Arc_{not} = Arc_O^{not} \cup Arc_I^{not}$, $Arc_O^{not} = \{(P, T), (T, P), (T, A_y)\}$, $Arc_I^{not} = \{(R_x, T)\}$,
- $w_d(a) = 1$ for all $a \in Arc_{not}$.

A part of implementation of the NOT gate model in the *Physarum* language (place-transition-arc mode) has the form:

```
#PETRI_NET
$setMode(PN.PTA);
P=new PN.Place("P");
P.setRole(PN.PHYSARUM);
Rx=new PN.Place("Rx");
Rx.setRole(PN.CONTROL_STIMULUS);
Ay=new PN.Place("Ay");
Ay.setRole(PN.OUTPUT_STIMULUS);
T=new PN.Transition("T");
```

Table 7.1 Main prototypes, corresponding to Petri net models, defined in the *Physarum* language, and their selected methods for the place-transition-arc mode

Prototype	Selected methods
PN.Place	*setDescription, setRole*
PN.Transition	*setDescription*
PN.Arc	*setAsInhibitor, setAsBidirectional*

Table 7.2 Main prototypes, corresponding to Petri net models, defined in the *Physarum* language, and their selected methods for the transition input/output mode

Prototype	Selected methods
PN.Place	*setDescription, setRole*
PN.Transition	*setDescription, addOrdinaryInput, addInhibitorInput, addOutput*

```
a1=new PN.Arc(P,T);
a1.setAsBidirectional;
a2=new PN.Arc(Rx,T);
a2.setAsInhibitor;
a3=new PN.Arc(T,Ay);
```

One can see that, for the transition $T \in Tr_{not}$, we can determine the following sets:

- $I_O(t) = \{P\}$,
- $I_I(t) = \{R_x\}$,
- $O(t) = \{P, A_y\}$.

A part of implementation of the NOT gate model in the *Physarum* language (transition input/output mode) has the form:

```
#PETRI_NET
$setMode(PN.IO);
P=new PN.Place("P");
P.setRole(PN.PHYSARUM);
Rx=new PN.Place("Rx");
Rx.setRole(PN.CONTROL_STIMULUS);
Ay=new PN.Place("Ay");
Ay.setRole(PN.OUTPUT_STIMULUS);
T=new PN.Transition("T");
T.addOrdinaryInput(P);
T.addInhibitorInput(Rx);
T.addOutput(Ay);
```

7.3   Physarum Language for Transition System Models

In case of transition system and timed transition system models, two modes are available in the *Physarum* language:

- state-event-transition mode,
- state precondition mode.

The main prototypes defined in the *Physarum* language and their selected methods are collected in Tables 7.3 and 7.4 for the state-event-transition mode and state pre-/postcondition mode, respectively.

Example 7.2 Let us consider a timed transition system $TTS = (S, E, T, S_{init}, l, u)$ given in Example 3.3. A part of implementation of TTS in the *Physarum* language (state-event-transition mode) has the form:

```
#TRANSITION_SYSTEM
$setMode(TS.SET);
s1=new TS.State("s1");
s1.setAsInitial;
s2=new TS.State("s2");
s3=new TS.State("s3");
s4=new TS.State("s4");
s5=new TS.State("s5");
e1=new TS.Event("e1");
t1=new TS.Transition(s1,e1,s2);
e2=new TS.Event("e2");
e2.setTimingConstraints(5,10);
t2=new TS.Transition(s1,e2,s3);
e3=new TS.Event("e3");
t3=new TS.Transition(s1,e3,s4);
```

Table 7.3 Main prototypes, corresponding to transition system and timed transition system models, defined in the *Physarum* language, and their selected methods for the state-event-transition mode

Prototype	Selected methods
TS.State	*setDescription, setAsInitial*
TS.Event	*setDescription, setTimingConstraints*
TS.Transition	

Table 7.4 Main prototypes, corresponding to transition system and timed transition system models, defined in the *Physarum* language, and their selected methods for the state precondition mode

Prototype	Selected methods
TS.State	*setDescription, setAsInitial, addPrecondition*

```
e4=new TS.Event("e4");
t4=new TS.Transition(s2,e4,s5);
```

A part of implementation of *TTS* in the *Physarum* language (state precondition mode) has the form:

```
#TRANSITION_SYSTEM
$setMode(TS.PRE);
s1=new TS.State("s1");
s1.setAsInitial;
s2=new TS.State("s2");
s3=new TS.State("s3");
s4=new TS.State("s4");
s5=new TS.State("s5");
s2.addPrecondition(s1);
s3.addPrecondition(s1);
s4.addPrecondition(s1);
s5.addPrecondition(s2);
```

7.4 Physarum Language for Tree Structures

In case of building tree structures, the main prototypes defined in the *Physarum* language and their selected methods are collected in Tables 7.5.

Example 7.3 Let us consider a tree structure $\mathbf{T} = (T, R_<)$ given in Example 5.10. The formal description of this tree structure is as follows:

$$T = \{x_1, x_2, x_3, x_4, x_5, x_6, x_7, x_8, x_9, x_{10}, x_{11}, x_{12}, x_{13}\},$$

and

$$R_< = \{(x_1, x_2), (x_1, x_3), (x_1, x_4), (x_2, x_5), (x_3, x_6), (x_3, x_7), (x_4, x_8), (x_4, x_9),$$
$$(x_5, x_{10}), (x_5, x_{11}), (x_6, x_{12}), (x_7, x_{13})\}.$$

Hence, we obtain:

- $Succ(x_1) = \{x_2, x_3, x_4\},$

Table 7.5 Main prototypes, corresponding to tree structures, defined in the *Physarum* language, and their selected methods

Prototype	Selected methods
TREE.Element	*setDescription, setAsInitial, setSuccessor*

- $Succ(x_2) = \{x_5\}$,
- $Succ(x_3) = \{x_6, x_7\}$,
- $Succ(x_4) = \{x_8, x_9\}$,
- $Succ(x_5) = \{x_{10}, x_{11}\}$,
- $Succ(x_6) = \{x_{12}\}$,
- $Succ(x_7) = \{x_{13}\}$,
- $Succ(x_8) = Succ(x_9) = Succ(x_{10}) = Succ(x_{11}) = Succ(x_{12}) = Succ(x_{13}) = \emptyset$.

A part of implementation of *TTS* in the *Physarum* language (state-event-transition mode) has the form:

```
#TREE_STRUCTURE
x1=new TREE.Element("x1");
x1.setAsInitial;
x2=new TREE.Element("x2");
x3=new TREE.Element("x3");
x4=new TREE.Element("x4");
x5=new TREE.Element("x5");
x6=new TREE.Element("x6");
x7=new TREE.Element("x7");
x8=new TREE.Element("x8");
x9=new TREE.Element("x9");
x10=new TREE.Element("x10");
x11=new TREE.Element("x11");
x12=new TREE.Element("x12");
x13=new TREE.Element("x13");
x1.setSuccessor(x2);
x1.setSuccessor(x3);
x1.setSuccessor(x4);
x2.setSuccessor(x5);
x3.setSuccessor(x6);
x3.setSuccessor(x7);
x4.setSuccessor(x8);
x4.setSuccessor(x9);
x5.setSuccessor(x10);
x5.setSuccessor(x11);
x6.setSuccessor(x12);
x7.setSuccessor(x13);
```

To sum up, our *Physarum* language can define programs simulating different *Physarum* machines: Petri nets, transition systems, and tree structures. Also, in this language we can define programs simulating *p*-adic arithmetic gates in *Physarum* machines (Chap. 9) and bio-inspired games (Chaps. 11 and 12).

Chapter 8
p-Adic Valued Logic

To implement arithmetic circuits on plasmodia we face the problem that the plasmodium is propagated in many directions simultaneously in accordance with stimuli and their topology. So, to manage this behaviour we need to limit possible ways of propagation by a number $p - 1$ of attractants for each original point of the plasmodium and for each next step of its transitions. In this case, we can interpret the plasmodium motion as the way of generating p-adic integers.

Let us remember that any p-adic integer has the following expansion [16, 50, 52]:

$$\beta_0 + \beta_1 \cdot p + \cdots + \beta_n \cdot p^n + \cdots = \sum_{n=0}^{\infty} \beta_n \cdot p^n,$$

where $\beta_n \in \{0, 1, \ldots, p - 1\}$, $\forall n \in \mathbf{N}$. This number sometimes has the following notation:

$$\ldots \beta_n \ldots \beta_3 \beta_2 \beta_1 \beta_0.$$

The set of all p-adic integers is denoted by $\mathbf{Z}_p$. Usual denary operations ($/$, $+$, $-$, $\cdot$) can be extrapolated to the case of them. For example, for 5-adic integers $\ldots 02324$ and $\ldots 003$ we obtain: $\frac{\ldots 02324}{\ldots 003} = \ldots 0423$ (the operation of division is not defined for all p-adic integers); $\ldots 02324 + \ldots 003 = \ldots 02332$; $\ldots 02324 - \ldots 003 = \ldots 02321$; $\ldots 02324 \cdot \ldots 003 = \ldots 013032$. Finite numbers of $\mathbf{Z}_p$ can be regarded as positive integers. So, we can identify $\ldots 02324$ with 339 and $\ldots 003$ with 3.

If $n \in \mathbf{Z}_p$, $n \neq 0$, and its canonical expansion contains only a finite number of nonzero digits α_j, then n is natural number (and vice versa). But if $n \in \mathbf{Z}_p$ and its expansion contains an infinite number of nonzero digits α_j, then n is an infinitely large natural number. Thus the set of p-adic integers contains actual infinities $n \in \mathbf{Z}_p \backslash \mathbf{N}$, $n \neq 0$. This is one of the most important features of non-Archimedean number

© Springer International Publishing AG, part of Springer Nature 2019
A. Schumann and K. Pancerz, *High-Level Models of Unconventional Computations*, Studies in Systems, Decision and Control 159,
https://doi.org/10.1007/978-3-319-91773-3_8

systems, therefore it is natural to compare $\mathbf{Z}_p$ with the set $^*\mathbf{N}$ of nonstandard natural numbers (hypernatural numbers).

Define a norm $|\cdot|_p \colon \mathbf{Q}_p \to \mathbf{R}$ on $\mathbf{Q}_p$ as follows:

$$\left| n = \sum_{k=N}^{\infty} \alpha_k \cdot p^k \right|_p :: = p^{-N},$$

where N is an index of the first number distinct from zero in p-adic expansion of n.

Note that $|0|_p :: = 0$. The function $|\cdot|_p$ has values 0 and $\{p^\gamma\}_{\gamma \in \mathbf{Z}}$ on $\mathbf{Q}_p$. Finally, $|x|_p \geq 0$ and $|x|_p = 0 \equiv x = 0$.

The function $\rho_p(x, y) = |x - y|_p$ is a metric on $\mathbf{Q}_p$. It is a translation invariant metric, i.e. $\rho_p(x + h, y + h) = \rho_p(x, y)$. As usual in metric spaces we define closed and open balls in $\mathbf{Q}_p$:

$$B[a, p^\gamma] :: = B_\gamma[a] = \{x \in \mathbf{Q}_p \colon \rho_p(x, a) \leq p^\gamma\},$$

$$B(a, p^\gamma) :: = B_\gamma(a) = \{x \in \mathbf{Q}_p \colon \rho_p(x, a) < p^\gamma\},$$

where $p^\gamma \in \mathbf{R}_+$.

The metric ρ_p satisfies the strong triangle inequality:

$$\rho_p(x, y) \leq \max(\rho_p(x, z), \rho_p(z, y)).$$

Such a kind of metric is called an *ultrametric* or *non-Archimedean metric*. We note that any open or closed ball in an ultrametric space is a simultaneously closed and open subset, because $B_{\gamma-1}[a] = B_\gamma(a)$.

The balls $B_\gamma(0)$ are additive subgroups of $\mathbf{Q}_p$: if $|x|_p, |y|_p \leq r$, then $|x + y|_p \leq \max(|x|_p, |y|_p) \leq r$. Furthermore, the ball $B[0, 1]$ is a ring: if $|x|_p, |y|_p \leq 1$, then $|x \cdot y|_p \leq |x|_p \cdot |y|_p \leq 1$. Moreover, the ball $B[0, 1]$ is the ring of p-adic integers $\mathbf{Z}_p$.

8.1   *p*-Adic Valued Matrix Logic

Extendthe standard order structure on $\{0, \ldots, p - 1\}$ to a partial order structure on $\mathbf{Z}_p$. Define this partial order structure on $\mathbf{Z}_p$ as follows:

$\mathcal{O}_{\mathbf{Z}_p}$ Let $x = \ldots x_n \ldots x_1 x_0$ and $y = \ldots y_n \ldots y_1 y_0$ be the canonical expansions of two p-adic integers $x, y \in \mathbf{Z}_p$. We set $x \leq y$ if we have $x_n \leq y_n$ for each $n = 0, 1, \ldots$ We set $x < y$ if we have $x_n \leq y_n$ for each $n = 0, 1, \ldots$ and there exists n_0 such that $x_{n_0} < y_{n_0}$. We set $x = y$ if $x_n = y_n$ for each $n = 0, 1, \ldots$

Now introduce two operations max, min in the partial order structure on $\mathbf{Z}_p$:

(1) for all p-adic integers $x, y \in \mathbf{Z}_p$, $\min(x, y) = x$ if and only if $x \leq y$ under condition $\mathcal{O}_{\mathbf{Z}_p}$,

(2) for all *p*-adic integers $x, y \in \mathbf{Z}_p$, $\max(x, y) = y$ if and only if $x \leq y$ under condition $\mathscr{O}_{\mathbf{Z}_p}$,

(3) for all *p*-adic integers $x, y \in \mathbf{Z}_p$, $\max(x, y) = \min(x, y) = x = y$ if and only if $x = y$ under condition $\mathscr{O}_{\mathbf{Z}_p}$.

The ordering relation $\mathscr{O}_{\mathbf{Z}_p}$ is not linear, but partial, because there exist elements $x, z \in \mathbf{Z}_p$, which are incompatible. As an example, let $p = 2$ and let $x = -\frac{1}{3} = \ldots 10101 \ldots 101$, $z = -\frac{2}{3} = \ldots 01010 \ldots 010$. Then the numbers x and z are incompatible.

Thus,

(4) Let $x = \ldots x_n \ldots x_1 x_0$ and $y = \ldots y_n \ldots y_1 y_0$ be the canonical expansions of two *p*-adic integers $x, y \in \mathbf{Z}_p$ and x, y are incompatible under condition $\mathscr{O}_{\mathbf{Z}_p}$. We get $\min(x, y) = z = \ldots z_n \ldots z_1 z_0$, where, for each $n = 0, 1, \ldots$, we set

   1. $z_n = y_n$ if $x_n \geq y_n$,
   2. $z_n = x_n$ if $x_n \leq y_n$,
   3. $z_n = x_n = y_n$ if $x_n = y_n$.

We get $\max(x, y) = z = \ldots z_n \ldots z_1 z_0$, where, for each $n = 0, 1, \ldots$, we set

   1. $z_n = y_n$ if $x_n \leq y_n$,
   2. $z_n = x_n$ if $x_n \geq y_n$,
   3. $z_n = x_n = y_n$ if $x_n = y_n$.

It is important to remark that there exists the maximal number $N_{max} \in \mathbf{Z}_p$ under condition $\mathscr{O}_{\mathbf{Z}_p}$. It is easy to see:

$$N_{max} = -1 = (p-1) + (p-1) \cdot p + \cdots + (p-1) \cdot p^k + \cdots = \sum_{k=0}^{\infty} (p-1) \cdot p^k$$

Therefore

(5) $\min(x, N_{max}) = x$ and $\max(x, N_{max}) = N_{max}$ for any $x \in \mathbf{Z}_p$.

Recall that a *matrix*, or *matrix logic*, $\mathfrak{M}$ for a language $\mathscr{L}$ is given by:

1. a nonempty set of truth values V of cardinality $|V| = m$,
2. a subset $D \subseteq V$ of designated truth values,
3. an algebra with domain V of appropriate type: for every *n*-place connective $\square$ of $\mathscr{L}$, there is an associated truth function $f \colon V^n \to V$, and
4. for every quantifier Q, there is an associated truth function $\widetilde{Q} \colon \wp(V) \backslash \emptyset \to V$

Notice that a truth function for quantifiers is a mapping from nonempty sets of truth values to truth values. For a non-empty set $M \subseteq V$, a quantified formula $Qx\varphi(x)$ takes the truth value $\widetilde{Q}(M)$ if, for every truth value $v \in V$, it holds that $v \in M$ iff there is a domain element d such that the truth value of φ in this point d is v (all relative to some interpretation). The set M is called the distribution of φ. For example, suppose that there are only two quantifiers in $\mathscr{L}$: the universal quantifier $\forall$ and the existential quantifier $\exists$. Further, we have the set of truth values $V = \{\top, \bot\}$, where $\bot$ is false and $\top$ is true, i.e. the set of designated truth values $D = \{\top\}$. Then we define the truth functions for the quantifiers $\forall$ and $\exists$ as follows:

1. $\widetilde{\forall}(\{\top\}) = \top$
2. $\widetilde{\forall}(\{\top, \bot\}) = \widetilde{\forall}(\{\bot\}) = \bot$
3. $\widetilde{\exists}(\{\bot\}) = \bot$
4. $\widetilde{\exists}(\{\top, \bot\}) = \widetilde{\exists}(\{\top\}) = \top$

Also, a matrix logic $\mathfrak{M}$ for a language $\mathscr{L}$ is an algebraic system $\mathfrak{M} = (V, f_0, f_1, \ldots, f_r, \widetilde{Q}_0, \widetilde{Q}_1, \ldots, \widetilde{Q}_q, D)$, where

1. V is a nonempty set of truth values for well-formed formulas of $\mathscr{L}$,
2. $f_0, f_1, \ldots, f_r$ are a set of matrix operations defined on the set V and assigned to corresponding propositional connectives $\square_0^{n_0}, \square_1^{n_1}, \ldots, \square_r^{n_r}$ of $\mathscr{L}$,
3. $\widetilde{Q}_0, \widetilde{Q}_1, \ldots, \widetilde{Q}_q$ are a set of matrix operations defined on the set V and assigned to corresponding quantifiers $Q_0, Q_1, \ldots, Q_q$ of $\mathscr{L}$,
4. D is a set of designated truth values such that $D \subseteq V$.

Now consider *p-adic valued matrix logic* $\mathfrak{M}_{\mathbf{Z}_p}$.

Definition 8.1 The ordered system $(V_{\mathbf{Z}_p}, \neg, \supset, \vee, \wedge, \widetilde{\exists}, \widetilde{\forall}, \{N_{max}\})$ is called *p-adic valued matrix logic* $\mathfrak{M}_{\mathbf{Z}_p}$, where

1. $V_{\mathbf{Z}_p} = \{0, \ldots, N_{max}\} = \mathbf{Z}_p$,
2. for all $x \in V_{\mathbf{Z}_p}$, $\neg x = N_{max} - x$,
3. for all $x, y \in V_{\mathbf{Z}_p}$, $x \supset y = (N_{max} - \max(x, y) + y)$,
4. for all $x, y \in V_{\mathbf{Z}_p}$, $x \vee y = (x \supset y) \supset y = \max(x, y)$,
5. for all $x, y \in V_{\mathbf{Z}_p}$, $x \wedge y = \neg(\neg x \vee \neg y) = \min(x, y)$,
6. for a subset $M \subseteq V_{\mathbf{Z}_p}$, $\widetilde{\exists}(M) = \max(M)$, where $\max(M)$ is a maximal element of M,
7. for a subset $M \subseteq V_{\mathbf{Z}_p}$, $\widetilde{\forall}(M) = \min(M)$, where $\min(M)$ is a minimal element of M,
8. $\{N_{max}\}$ is the set of designated truth values.

The truth value $0 \in \mathbf{Z}_p$ is false, the truth value $N_{max} \in \mathbf{Z}_p$ is true, and other truth values $x \in \mathbf{Z}_p$ are neutral.

Proposition 8.1 *The logic* $\mathfrak{M}_{\mathbf{Z}_2} = (V_{\mathbf{Z}_2}, \neg, \supset, \vee, \wedge, \widetilde{\exists}, \widetilde{\forall}, \{N_{max}\})$ *is a Boolean algebra.*

Proof Indeed, the operation $\neg$ in $\mathfrak{M}_{\mathbf{Z}_2}$ is the Boolean complement:

1. $\max(x, \neg x) = N_{max}$,
2. $\min(x, \neg x) = 0$. $\square$

A nonempty subset of truth values $\Delta \subseteq V_{\mathbf{Z}_p}$ is said to be an *ideal* on truth values of $\mathfrak{M}_{\mathbf{Z}_p}$ if the following condition holds:
for all $x, y \in \mathbf{Z}_p$, $\max(x, y) \in \Delta$ iff $x \in \Delta$ and $y \in \Delta$.

Proposition 8.2 *Suppose an ideal Δ in $\mathfrak{M}_{\mathbf{Z}_p}$ has a maximal element* $\max(\Delta) \in \Delta$. *Then Δ is the subset of the closed ball* $B_\gamma[0] = \{x \in \mathbf{Q}_p : \rho_p(x, 0) \leq p^\gamma\}$, *where* $p^\gamma = |\max(\Delta)|_p$.

Proof Give an example. Let $\alpha = \dots 0\alpha_3\alpha_2 00$, $\beta = \dots \beta_4 0000$ be the *p*-adic integers and $\alpha, \beta \in \Delta$. Then $\max(\alpha, \beta) = \dots \beta_4\alpha_3\alpha_2 00$ and $\max(\alpha, \beta) \in \Delta$. We obtain: $|\max(\alpha, \beta)|_p = p^{-2}$ and $|\alpha|_p \leq p^{-2}$, $|\beta|_p \leq p^{-2}$. $\qquad\square$

A subset of truth values $\Delta \subseteq V_{\mathbf{Z}_p}$ is called a *complete ideal* if Δ is defined as follows:

1. Δ is an ideal,
2. $\Delta \neq \{0\}$,
3. if N is a maximal truth value that is contained in Δ, then Δ contains all truth values that are smaller than N.

Proposition 8.3 *Suppose a complete ideal Δ in $\mathfrak{M}_{\mathbf{Z}_p}$ has a maximal element N_{max}. Then Δ is the closed ball* $B[0, 1] = \{x \in \mathbf{Q}_p : \rho_p(x, 0) \leq 1\}$.

Proof It follows from Proposition 8.2. $\qquad\square$

About *p*-adic valued logics please see [49, 77–82, 84]. About their implementations in the slime mould behaviour see [76, 89, 94, 95, 103, 106, 109].

8.2 *p*-Adic Valued Propositional Logical Language

Definition 8.2 A *p*-adic valued propositional logical language $\mathscr{L}_{\mathbf{Z}_p}^\infty$ consists of the following symbols:

1. First-order propositional formulas of *p*-valued Łukasiewicz's propositional logic $Ł_p$ (about *p*-valued logic and its deductive system please see [15]): $\varphi, \phi, \psi, \dots$(we suppose implicitly that $Ł_p$ is truth-functionally complete thanks to Słupecki's operators (constant maps) $T_k(\phi) : v(\phi) \rightarrow k \in \{1, \dots, p - 2\}$, where $v(\phi)$ is any truth valuation of $\phi \in Ł_p$).
2. Logical symbols:

    (i) Various order propositional connectives of arity n_j: $\Box_0^{n_0}, \Box_1^{n_1}, \dots, \Box_r^{n_r}$, which are built by superposition of negation $\neg$ and implication $\supset$.

(ii) Vertical quantifiers of various order: $\forall_1^0, \ldots, \forall_1^{p-1}, \exists_1^0, \ldots, \exists_1^{p-1}, \forall_2^0, \ldots,$
$\forall_2^{p-1}, \exists_2^0, \ldots, \exists_2^{p-1}, \ldots, \forall_{i-1}^0, \ldots, \forall_{i-1}^{p-1}, \exists_{i-1}^0, \ldots, \exists_{i-1}^{p-1}, \ldots$, where the
order is denoted by the lower index of quantifiers: an i-order quantifier
has the lower index $i - 1$.

3. Auxiliary symbols: (,), and , (comma).

Definition 8.3 Well-formed formulas of $\mathscr{L}_{\mathbb{Z}_p}^\infty$ are inductively defined as follows:

1. If φ is a first-order propositional formula and $Q_1, Q_2, \ldots, Q_{i-1}$ are a finite
sequence of vertical quantifiers, then

$$Q_{i-1}(\ldots(Q_1\varphi(\varphi))\ldots)\ldots(Q_1\varphi(\varphi))$$

is an i-order formula denoted sometimes by φ_i. It is called atomic or an atom. Its
outermost logical symbols are $Q_1, Q_2, \ldots, Q_{i-1} \in \{\forall_1^0, \ldots, \forall_1^{p-1}, \exists_1^0, \ldots, \exists_1^{p-1},$
$\forall_2^0, \ldots, \forall_2^{p-1}, \exists_2^0, \ldots, \exists_2^{p-1}, \ldots, \forall_{i-1}^0, \ldots, \forall_{i-1}^{p-1}, \exists_{i-1}^0, \ldots, \exists_{i-1}^{p-1}\}$. If the first of
these quantifiers is $\forall_1^{y_1}$, then the other are also universal with the upper indices
that are equal to y_1. If the first of these quantifiers is $\exists_1^{y_1}$, then the other are also
existential with the upper indices that aren't possibly equal to y_1.
2. If $\varphi_i, \ldots, \psi_j$ are formulas of various order (i.e. φ_i is an i-order formula, ψ_j is
a j-order formula, etc.) and $\square^n$ is a propositional connective of arity n, then
$\square^n(\varphi_i, \ldots, \psi_j)$ is a formula with outermost logical symbol $\square^n$ and this formula
is k-order in the case $\max(i, \ldots, j) = k$. In other words, the order of formulas
that contain propositional connectives is the maximal order of atoms.
3. If φ is a first-order propositional formula and $Q_1, Q_2, \ldots, Q_{i-1}, \ldots$ are an infinite
sequence of vertical quantifiers, then

$$\ldots Q_{i-1}(\ldots(Q_1\varphi(\varphi))\ldots)\ldots(Q_1\varphi(\varphi))$$

is an infinite-order formula denoted sometimes by φ_∞. It is called atomic or an
atom. Its outermost logical symbols are $Q_1, Q_2, \ldots, Q_{i-1}, \ldots \in \{\forall_1^0, \ldots, \forall_1^{p-1},$
$\exists_1^0, \ldots, \exists_1^{p-1}, \forall_2^0, \ldots, \forall_2^{p-1}, \exists_2^0, \ldots, \exists_2^{p-1}, \ldots, \forall_{i-1}^0, \ldots, \forall_{i-1}^{p-1}, \exists_{i-1}^0, \ldots, \exists_{i-1}^{p-1},$
$\ldots\}$. If the first of these quantifiers is $\forall_1^{y_1}$, then the other are also universal with
the upper indices that are equal to y_1. If the first of these quantifiers is $\exists_1^{y_1}$, then the
other are also existential with the upper indices that aren't possibly equal to y_1.
4. If $\varphi_\infty, \ldots, \psi_\infty$ are formulas of infinite order and $\square^n$ is a propositional connective
of arity n, then $\square^n(\varphi_\infty, \ldots, \psi_\infty)$ is a formula with outermost logical symbol $\square^n$
and this formula is infinite-order.

A quantifier of the form $Q_j^{y_j \in \{0, \ldots, p-1\}}(\ldots(\varphi))\ldots(\varphi)$, where $Q_j^{y_j \in \{0, \ldots, p-1\}} \in$
$\{\forall_j^0, \ldots, \forall_j^{p-1}, \exists_j^0, \ldots, \exists_j^{p-1}\}$, $j = 1, 2, \ldots$, is a vertical quantifier. Vertical quantifiers satisfy the following condition:

$$\neg Q_{i-1}^{y_{i-1}\in\{0,\dots,p-1\}}(Q_{i-2}^{y_{i-2}\in\{0,\dots,p-1\}}(\dots(Q_1^{y_1\in\{0,\dots,p-1\}}\varphi(\varphi)))\dots)\dots$$
$$(Q_1^{y_1\in\{0,\dots,p-1\}}\varphi(\varphi))\equiv$$
$$Q_{i-1}'^{(p-1)-y_{i-1}}(Q_{i-2}'^{(p-1)-y_{i-2}}(\dots(Q_1'^{(p-1)-y_1}\varphi(\neg\varphi)))\dots)\dots$$
$$(Q_1'^{(p-1)-y_1}\varphi(\neg\varphi)),$$

where $Q_j^{y_j\in\{0,\dots,p-1\}}, Q_j'^{(p-1)-y_j} \in \{\forall_j^0,\dots,\forall_j^{p-1}, \exists_j^0,\dots,\exists_j^{p-1}\}$, $j=1,2,\dots$, i.e. we replace a quantifier $\forall$ (respectively, $\exists$) by its dual quantifier $\exists$ (respectively, $\forall$) when we relocate the negation to the propositional expression.

Let us remember that there is a sequence of finite-valued Łukasiewicz's logics Ł$_2$, Ł$_3$, Ł$_5$, Ł$_7$, …At the same time, we can obtain for all logics an infinite-order extension such that only *p*-adic integers with respect to the norms $|\cdot|_2$, $|\cdot|_3$, $|\cdot|_5$, $|\cdot|_7$,…are the range of values for their well-formed formulas. In other words, we can construct a *p*-adic valued propositional logical language $\mathscr{L}_{\mathbf{Z}_p}^\infty$ for all logics Ł$_2$, Ł$_3$, Ł$_5$, Ł$_7$, …(see Table 8.1).

Consider a 2-valued first-order propositional logic Ł$_2$, which has the set of truth values V such that $|V|=2$. Then vertical quantifiers in $\mathscr{L}_{\mathbf{Z}_2}^\infty$ have the following natural many-valued interpretation:

1. The universal vertical quantifier $\forall_1^1\varphi(\varphi)$ means that a first-order formula φ has the truth value 1 for all its interpretations. Therefore a second-order formula $\forall_1^1\varphi(\varphi)$ is true (it has the truth value 3) if a first-order formula φ is a tautology and $\forall_1^1\varphi(\varphi)$ is not true otherwise, e.g. it is false (it has the truth value 0) if φ has the truth value 0 for some its interpretations. A second-order formula $\forall_1^1\varphi(\varphi)$ has the truth value 1 if a first-order formula φ has the truth value 0 for all its interpretations. A second-order formula $\forall_1^1\varphi(\varphi)$ has the truth value 2 if a first-order formula φ has the truth value 1 for some its interpretations.
2. The universal vertical quantifier $\forall_1^0\varphi(\varphi)$ means that a first-order formula φ has the truth value 0 for all its interpretations. Therefore a second-order formula $\forall_1^0\varphi(\varphi)$ is true (it has the truth value 3) if φ is a contradiction. A second-order formula $\forall_1^0\varphi(\varphi)$ has the truth value 0 if a first-order formula φ has the truth value 1 for some its interpretations. A second-order formula $\forall_1^0\varphi(\varphi)$ has the truth value 1 if

Table 8.1 The infinite hierarchy of *p*-adic valued logics

	1-st prime number (2)	2-nd prime number (3)	3-rd prime number (5)	…	*s*-th prime number (p_s)
1-st order logical language	$\varphi_1 \in \mathscr{L}_{\mathbf{Z}_2}^1$	$\varphi_1 \in \mathscr{L}_{\mathbf{Z}_3}^1$	$\varphi_1 \in \mathscr{L}_{\mathbf{Z}_5}^1$	…	$\varphi_1 \in \mathscr{L}_{\mathbf{Z}_{p_s}}^1$
2-nd order logical language	$\varphi_2 \in \mathscr{L}_{\mathbf{Z}_2}^2$	$\varphi_2 \in \mathscr{L}_{\mathbf{Z}_3}^2$	$\varphi_2 \in \mathscr{L}_{\mathbf{Z}_5}^2$	…	$\varphi_2 \in \mathscr{L}_{\mathbf{Z}_{p_s}}^2$
3-rd order logical language	$\varphi_3 \in \mathscr{L}_{\mathbf{Z}_2}^3$	$\varphi_3 \in \mathscr{L}_{\mathbf{Z}_3}^3$	$\varphi_3 \in \mathscr{L}_{\mathbf{Z}_5}^3$	…	$\varphi_3 \in \mathscr{L}_{\mathbf{Z}_{p_s}}^3$
$\vdots$	$\vdots$	$\vdots$	$\vdots$	$\vdots$	$\vdots$
∞-order logical language	$\varphi_\infty \in \mathscr{L}_{\mathbf{Z}_2}^\infty$	$\varphi_\infty \in \mathscr{L}_{\mathbf{Z}_3}^\infty$	$\varphi_\infty \in \mathscr{L}_{\mathbf{Z}_5}^\infty$	…	$\varphi_\infty \in \mathscr{L}_{\mathbf{Z}_{p_s}}^\infty$

a first-order formula φ has the truth value 1 for all its interpretations. A second-order formula $\forall_1^0 \varphi(\varphi)$ has the truth value 2 if a first-order formula φ has the truth value 0 for some its interpretations.

3. The universal vertical quantifier $\forall_2^1 (\forall_1^1 \varphi(\varphi))(\forall_1^1 \varphi(\varphi))$ means that a second-order atomic formula φ_2 has the truth value 3 for all its interpretations.

4. The universal vertical quantifier $\forall_2^0 (\forall_1^0 \varphi(\varphi))(\forall_1^0 \varphi(\varphi))$ means that a second-order atomic formula φ_2 has the truth value 0 for all its interpretations.

5. The universal vertical quantifier $\forall_2^0 (\forall_1^1 \varphi(\varphi))(\forall_1^1 \varphi(\varphi))$ means that a second-order atomic formula φ_2 has the truth value 1 for all its interpretations.

6. The universal vertical quantifier $\forall_2^1 (\forall_1^0 \varphi(\varphi))(\forall_1^0 \varphi(\varphi))$ means that a second-order atomic formula φ_2 has the truth value 2 for all its interpretations.

7. The existential vertical quantifier $\exists_1^1 \varphi(\varphi)$ means that a first-order formula φ has the truth value 1 for some its interpretations. If this condition holds, then a second-order formula $\exists_1^1 \varphi(\varphi)$ has the truth value 3. A second-order formula $\exists_1^1 \varphi(\varphi)$ has the truth value 0 if a first-order formula φ has the truth value 0 for all its interpretations. A second-order formula $\exists_1^1 \varphi(\varphi)$ has the truth value 1 if a first-order formula φ has the truth value 0 for some its interpretations. A second-order formula $\exists_1^1 \varphi(\varphi)$ has the truth value 2 if a first-order formula φ has the truth value 1 for all its interpretations.

8. The existential vertical quantifier $\exists_1^0 \varphi(\varphi)$ means that a first-order formula φ has the truth value 0 for some its interpretations. In this case a second-order formula $\exists_1^0 \varphi(\varphi)$ has the truth value 3. A second-order formula $\exists_1^0 \varphi(\varphi)$ has the truth value 0 if a first-order formula φ has the truth value 1 for all its interpretations. A second-order formula $\exists_1^0 \varphi(\varphi)$ has the truth value 1 if a first-order formula φ has the truth value 1 for some its interpretations. A second-order formula $\exists_1^0 \varphi(\varphi)$ has the truth value 2 if a first-order formula φ has the truth value 0 for all its interpretations.

9. The existential vertical quantifier

$$\exists_2^1 (\exists_1^{y_1 \in \{0,1\}} \varphi(\varphi))(\exists_1^{y_1 \in \{0,1\}} \varphi(\varphi))$$

means that a second-order atomic formula φ_2 has the truth value $\alpha \in \{2, 3\}$ for some its interpretations. We shall say that the satisfiability degree of these interpretations for α is 1.

10. The existential vertical quantifier

$$\exists_2^0 (\exists_1^{y_1 \in \{0,1\}} \varphi(\varphi))(\exists_1^{y_1 \in \{0,1\}} \varphi(\varphi))$$

means that a second-order atomic formula φ_2 has the truth value $\alpha' \in \{0, 1\}$ for some its interpretations. We shall say that the satisfiability degree of these interpretations for $\alpha = \neg\alpha'$ is 0.

11. The universal vertical quantifier

$$\forall_{i-1}^{y_{i-1} \in \{0,1\}} (\forall_{i-2}^{y_{i-2} \in \{0,1\}} (\ldots (\forall_1^{y_1 \in \{0,1\}} \varphi(\varphi))) \ldots) \ldots (\forall_1^{y_1 \in \{0,1\}} \varphi(\varphi))$$

means that an $(i-1)$-order atomic formula

$$\forall_{i-2}^{y_{i-2}\in\{0,1\}}(\dots(\forall_1^{y_1\in\{0,1\}}\varphi(\varphi)))\dots(\forall_1^{y_1\in\{0,1\}}\varphi(\varphi))$$

has the truth value $\alpha = (\sum_{k=1}^{i-2} y_k \cdot 2^{k-1})$ for all its interpretations. Therefore

a. an *i*-order atomic formula $\varphi_i = \forall_{i-1}^{y_{i-1}\in\{0,1\}}(\dots(\varphi))\dots(\varphi)$ has the truth value $\underbrace{1\dots1}_{i}$ if this condition holds;

b. an *i*-order atomic formula φ_i has the truth value 0 if an $(i-1)$-order atomic formula φ_{i-1} has the truth value $\alpha' = (\sum_{k=1}^{i-2} \alpha'_k \cdot 2^{k-1})$ for some its interpretations, where $\alpha'_k = 1 - y_k$ for every $k \in \{1,\dots,i-2\}$;

c. an *i*-order atomic formula φ_i has the truth value $\underbrace{0\dots0}_{i-1}1$ if an $(i-1)$-order atomic formula φ_{i-1} has the truth value $\alpha' = (\sum_{k=1}^{i-2} \alpha'_k \cdot 2^{k-1})$ for all its interpretations, where $\alpha'_k = 1 - y_k$ for every $k \in \{1,\dots,i-2\}$, etc.

12. The existential vertical quantifier

$$\exists_{i-1}^{y_{i-1}\in\{0,1\}}(\exists_{i-2}^{y_{i-2}\in\{0,1\}}(\dots(\exists_1^{y_1\in\{0,1\}}\varphi(\varphi)))\dots)\dots(\exists_1^{y_1\in\{0,1\}}\varphi(\varphi))$$

means that an $(i-1)$-order atomic formula

$$\exists_{i-2}^{y_{i-2}\in\{0,1\}}(\dots(\exists_1^{y_1\in\{0,1\}}\varphi(\varphi)))\dots(\exists_1^{y_1\in\{0,1\}}\varphi(\varphi))$$

has the truth value $\alpha = (\sum_{k=1}^{i-2} y_{k-1} \cdot 2^k)$ for some its interpretations and the satisfiability degree of these interpretations for α is y_{i-1}. Therefore

a. an *i*-order atomic formula $\varphi_i = \exists_{i-1}^{y_{i-1}\in\{0,1\}}(\dots(\varphi))\dots(\varphi)$ has the truth value $\underbrace{1\dots1}_{i}$ if this condition holds;

b. an *i*-order atomic formula φ_i has the truth value $\underbrace{0\dots0}_{i}$ if an $(i-1)$-order atomic formula φ_{i-1} has the truth value $\alpha' = (2^{i-1} - 1) - \alpha = (\sum_{k=1}^{i-2}(1 - y_k) \cdot 2^{k-1})$ for all its interpretations;

c. an *i*-order atomic formula φ_i has the truth value $\underbrace{0\dots0}_{i-1}1$ if an $(i-1)$-order atomic formula φ_{i-1} has the truth value $\alpha' = (\sum_{k=1}^{i-2} \alpha'_k \cdot 2^{k-1})$ for some its interpretations, where $\alpha'_k = 1 - y_k$ for every $k \in \{1,\dots,i-2\}$, etc.

Getting the sequence of atomic formulas φ_∞, we go over from *n*-order language to $(n+1)$-order language. As a result we obtain the tree of atomic formulas, namely, if we take a binary logic $Ł_2$, then 'verum' or 'falsum' of its formulas will be defined at the first level, their 'general validity', 'satisfiability', 'non-satisfiability' or 'non-general validity' will be defined at the second level, etc. Notice that the truth value 'verum' corresponds to the sense of the 1-order formula φ, the truth value 'falsum' corresponds to the sense of the 1-order formula $\neg\varphi$, the truth value 'general validity' corresponds to the sense of the 2-order atom $\forall_1^1\varphi(\varphi)$, the truth value 'satisfiability' corresponds to the sense of the 2-order atom $\exists_1^1\varphi(\varphi)$, the truth value 'non-general validity' corresponds to the sense of the 2-order atom $\exists_1^0\varphi(\varphi)$, the truth value 'non-satisfiability' corresponds to the sense of the 2-order atom $\forall_1^0\varphi(\varphi)$, etc. In other words, we have a 2-valued logic at the first level and a 4-valued logic at the second level, etc.

This implies that if we build a matrix logic for all sequences φ_n beginning from expressions φ of Ł$_2$, then $D = \{(2^i - 1)_{i=1}^n\}$, where i is a level of extension, is the set of designated truth values, and $V = \{0, 1, 2, \ldots, (2^i - 1)_{i=1}^n\}$ is the set of all truth values. Consequently the infinite extension of binary logic Ł$_2$ has $D = \{(2^i - 1)_{i=1}^\infty\}$ as the set of designated truth values and it has $V = \{0, 1, 2, \ldots, (2^i - 1)_{i=1}^\infty\}$ as the set of all truth values.

Notice that the structure of infinite-order 2-adic propositional logical language $\mathscr{L}_{\mathbf{Z}_2}^\infty$ is a Boolean algebra (see Proposition 8.1).

Since

$$\neg\forall_i^0(\ldots(\varphi))\ldots(\varphi) = \exists_i^1(\ldots(\neg\varphi))\ldots(\neg\varphi),$$

$$\neg\exists_i^0(\ldots(\varphi))\ldots(\varphi) = \forall_i^1(\ldots(\neg\varphi))\ldots(\neg\varphi),$$

where $i = 1, 2, \ldots$, we see that we can consider only quantifiers $\forall_i^1$, $\exists_i^1$ and ignore the upper index of vertical quantifiers in 2-adic language. Therefore, in the second-order 2-adic propositional logical language, we will denote sometimes the truth value 0 by $\ulcorner\forall\varphi(\neg\varphi)\urcorner$, the truth value 1 by $\ulcorner\exists\varphi(\neg\varphi)\urcorner$, the truth value 2 by $\ulcorner\exists\varphi(\varphi)\urcorner$, the truth value 3 by $\ulcorner\forall\varphi(\varphi)\urcorner$.[1] In the third-order logic we will denote sometimes the truth value 0 by $\ulcorner\forall^0(\forall^0\varphi(\varphi))(\forall^0\varphi(\varphi))\urcorner$, the truth value 1 by $\ulcorner\forall^0(\forall^1\varphi(\varphi))(\forall^1\varphi(\varphi))\urcorner$, the truth value 2 by $\ulcorner\exists^0(\exists^0\varphi(\varphi))(\exists^0\varphi(\varphi))\urcorner$, the truth value 3 by $\ulcorner\exists^0(\exists^1\varphi(\varphi))(\exists^1\varphi(\varphi))\urcorner$, the truth value 4 by $\ulcorner\exists^1(\exists^0\varphi(\varphi))(\exists^0\varphi(\varphi))\urcorner$, the truth value 5 by $\ulcorner\exists^1(\exists^1\varphi(\varphi))(\exists^1\varphi(\varphi))\urcorner$, the truth value 6 by $\ulcorner\forall^1(\forall^0\varphi(\varphi))(\forall^0\varphi(\varphi))\urcorner$, the truth value 7 by $\ulcorner\forall^1(\forall^1\varphi(\varphi))(\forall^1\varphi(\varphi))\urcorner$, etc., see Fig. 8.1.

We can set truth values of higher-order formulas on the base of its first-order truth matrix. For example, an i-order formula φ_i has a truth value $a_i \in \{0, \ldots, 2^i - 1\}$ if the following conditions hold:

1. If a first-order propositional formula φ has the value 1 for all its interpretations I and φ contains n propositional variables, then, at the level $i = 2^n$, the expression

$$\forall_{i-1}(\forall_{i-2}(\ldots(\forall_1\varphi(\varphi)))\ldots)\ldots(\forall_1\varphi(\varphi))$$

---

[1] Logical operations over formulas $\forall\varphi(\varphi)$, $\exists\varphi(\varphi)$, $\exists\varphi(\neg\varphi)$, $\forall\varphi(\neg\varphi)$ of the second-order logic are closed for the set of truth values 0, 1, 2, 3, therefore any logical operation over the expressions $\ulcorner\forall\varphi(\varphi)\urcorner$, $\ulcorner\exists\varphi(\varphi)\urcorner$, $\ulcorner\exists\varphi(\neg\varphi)\urcorner$, $\ulcorner\forall\varphi(\neg\varphi)\urcorner$ is equivalent to one of them.

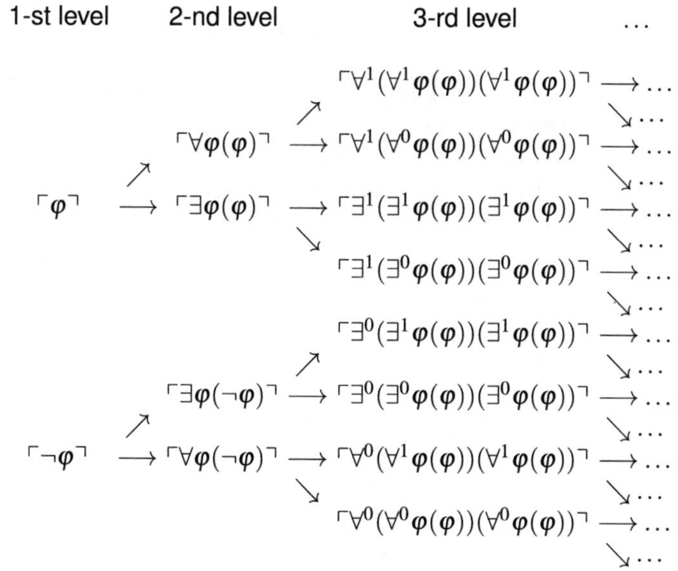

1-st level   2-nd level   3-rd level   . . .

$\ulcorner \forall^1 (\forall^1 \varphi(\varphi))(\forall^1 \varphi(\varphi)) \urcorner \longrightarrow \dots$

$\nearrow \qquad \searrow \dots$

$\ulcorner \forall \varphi(\varphi) \urcorner \longrightarrow \ulcorner \forall^1 (\forall^0 \varphi(\varphi))(\forall^0 \varphi(\varphi)) \urcorner \longrightarrow \dots$

$\nearrow \qquad \searrow \dots$

$\ulcorner \varphi \urcorner \longrightarrow \ulcorner \exists \varphi(\varphi) \urcorner \longrightarrow \ulcorner \exists^1 (\exists^1 \varphi(\varphi))(\exists^1 \varphi(\varphi)) \urcorner \longrightarrow \dots$

$\searrow \qquad \searrow \dots$

$\ulcorner \exists^1 (\exists^0 \varphi(\varphi))(\exists^0 \varphi(\varphi)) \urcorner \longrightarrow \dots$

$\searrow \dots$

$\ulcorner \exists^0 (\exists^1 \varphi(\varphi))(\exists^1 \varphi(\varphi)) \urcorner \longrightarrow \dots$

$\nearrow \qquad \searrow \dots$

$\ulcorner \exists \varphi(\neg\varphi) \urcorner \longrightarrow \ulcorner \exists^0 (\exists^0 \varphi(\varphi))(\exists^0 \varphi(\varphi)) \urcorner \longrightarrow \dots$

$\nearrow \qquad \searrow \dots$

$\ulcorner \neg\varphi \urcorner \longrightarrow \ulcorner \forall \varphi(\neg\varphi) \urcorner \longrightarrow \ulcorner \forall^0 (\forall^1 \varphi(\varphi))(\forall^1 \varphi(\varphi)) \urcorner \longrightarrow \dots$

$\searrow \qquad \searrow \dots$

$\ulcorner \forall^0 (\forall^0 \varphi(\varphi))(\forall^0 \varphi(\varphi)) \urcorner \longrightarrow \dots$

$\searrow \dots$

Fig. 8.1 If φ is a matrix (e.g., a Boolean function), then its complexity depends on the number n of variables ($n \to \infty$) in φ. In this case the set of values $V = \{0, 1, \dots, (2^i - 1)^\infty_{i=1}\}$ shows the degree of satisfiability or non-general validity of a formula φ for each level n ($n \to \infty$)

has the truth value $2^i - 1$ and the expression

$$\exists_{i-1}(\exists_{i-2}(\dots (\exists_1 \varphi(\neg\varphi)))\dots)\dots(\exists_1 \varphi(\neg\varphi))$$

has the truth value 0. The other expressions have truth values in the interval [1, $2^i - 2$]. According to Fig. 8.1, the natural meaning of $\forall^1_i (\forall^1_{i-1} \dots (\varphi))$ is greater than the natural meaning of $\forall^1_i (\forall^0_{i-1} \dots (\varphi))$, the meaning of $\forall^1_i (\forall^0_{i-1} \dots (\varphi))$ is greater than the meaning of $\exists^1_i (\exists^1_{i-1} \dots (\varphi))$, the meaning of $\exists^1_i (\exists^1_{i-1} \dots (\varphi))$ is greater than the meaning of $\exists^1_i (\exists^0_{i-1} \dots (\varphi))$, the meaning of $\exists^1_i (\exists^0_{i-1} \dots (\varphi))$ is greater than the meaning of $\exists^0_i (\exists^0_{i-1} \dots (\varphi))$, the meaning of $\exists^0_i (\exists^0_{i-1} \dots (\varphi))$ is greater than the meaning of $\forall^0_i (\forall^1_{i-1} \dots (\varphi))$, the meaning of $\forall^0_i (\forall^1_{i-1} \dots (\varphi))$ is greater than the meaning of $\forall^0_i (\forall^0_{i-1} \dots (\varphi))$ for $i = 1, 2, \dots$, etc.

2. If a formula φ has the value 0 for all its interpretations I and φ contains n propositional variables, then, at the level $i = 2^n$, the expression

$$\forall_{i-1}(\forall_{i-2}(\dots (\forall_1 \varphi(\neg\varphi)))\dots)\dots(\forall_1 \varphi(\neg\varphi))$$

has the truth value $2^i - 1$ and the expression

$$\exists_{i-1}(\exists_{i-2}(\dots (\exists_1 \varphi(\varphi)))\dots)\dots(\exists_1 \varphi(\varphi))$$

has the truth value 0. The other expressions have truth values in the interval [1, $2^i - 2$].

3. If a formula φ has the value 1 only for some its interpretations I and φ contains n propositional variables, then, for $i = 2^n$, its values can be defined as follows (just if we want to fix truth values of the first order).

Recall that there exists 2^n interpretations of φ: $I_1, I_2, \ldots I_{2^n}$. These interpretations ascribe matrix values to φ. There are 2^{2^n} combinations of truth values of φ:

$$
\begin{array}{ll}
0 & \underbrace{(00\ldots0)}_{2^n} \\[2mm]
1 & \underbrace{(00\ldots0\,1)}_{2^n-1} \\[2mm]
2 & \underbrace{(00\ldots0\,10)}_{2^n-2} \\[2mm]
3 & \underbrace{(00\ldots0\,11)}_{2^n-2} \\[2mm]
\vdots & \quad\vdots \\[2mm]
2^{2^n}-1 & \underbrace{(11\ldots1)}_{2^n}
\end{array}
$$

If $i = 2^n$, then there exist 2^{2^n} combinations of 2^n-order vertical existential quantifiers:

$$
\begin{array}{ll}
0 & \exists^0_{2^n-1}(\exists^0_{2^n-2}(\exists^0_{2^n-3}(\ldots(\exists^0_1\varphi(\varphi))))\ldots)\ldots(\exists^0_1\varphi(\varphi)) \\
1 & \exists^0_{2^n-1}(\exists^0_{2^n-2}(\exists^0_{2^n-3}(\ldots(\exists^1_1\varphi(\varphi))))\ldots)\ldots(\exists^1_1\varphi(\varphi)) \\
2 & \exists^0_{2^n-1}(\exists^0_{2^n-2}(\ldots(\exists^0_3(\exists^1_2(\exists^0_1\varphi(\varphi))))\ldots))\ldots(\exists^0_1\varphi(\varphi)) \\
3 & \exists^0_{2^n-1}(\exists^0_{2^n-2}(\ldots(\exists^0_3(\exists^1_2(\exists^1_1\varphi(\varphi))))\ldots))\ldots(\exists^1_1\varphi(\varphi)) \\
4 & \exists^0_{2^n-1}(\exists^0_{2^n-2}(\ldots(\exists^1_3(\exists^0_2(\exists^0_1\varphi(\varphi))))\ldots))\ldots(\exists^0_1\varphi(\varphi)) \\
\vdots & \quad\vdots \\
2^{2^n}-2 & \exists^1_{2^n-1}(\exists^1_{2^n-2}(\exists^1_{2^n-3}(\ldots(\exists^0_1\varphi(\varphi))))\ldots)\ldots(\exists^0_1\varphi(\varphi)) \\
2^{2^n}-1 & \exists^1_{2^n-1}(\exists^1_{2^n-2}(\exists^1_{2^n-3}(\ldots(\exists^1_1\varphi(\varphi))))\ldots)\ldots(\exists^1_1\varphi(\varphi))
\end{array}
$$

We see that there is the bijection, which takes the set of all combinations of matrix values for φ to the set of all 2^n-order formulas φ_{2^n}. A corresponding quantifier combination is a true expression for φ_{2^n} in the 2^{2^n}-valued logic.

Example 8.1 Let us consider an expression $\psi(x, y, z)$ that contains only 3 propositional variables. Evidently, there are 2^8 combinations of truth values for $\psi(x, y, z)$:

$$(0, 0, 0, 0, 0, 0, 0, 0), \ldots, (1, 1, 1, 1, 1, 1, 0, 1), (1, 1, 1, 1, 1, 1, 1, 0), (1, 1, 1, 1, 1, 1, 1, 1)$$

We assign one of the 2^8 combinations of 8-order existential quantifiers to each combination of truth values for $\psi(x, y, z)$.

Suppose, the original expression $\psi(x, y, z)$ has the truth table (matrix):

x	y	z	$\psi(x, y, z)$
1	1	1	1
1	1	0	0
1	0	1	0
1	0	0	0
0	1	1	0
0	1	0	0
0	0	1	0
0	0	0	0

Then it has the following true denoting in 8-order 2-adic propositional logical language:

$$\exists^1(\exists^0(\exists^0(\exists^0(\exists^0(\exists^0(\exists^0\psi(\psi)))))))\ldots(\exists^0(\exists^0\psi(\psi(x, y, z))))(\exists^0\psi(\psi(x, y, z))).$$

In the 8-order matrix logic $\mathfrak{M}_{\mathbf{Z}_2}$, this denoting corresponds to the truth value 2^{i-1} (where $i = 8$). The negation of this value is the expression

$$\ulcorner\exists^0(\exists^1(\exists^1(\exists^1(\exists^1(\exists^1(\exists^1\psi(\psi)))))))\ldots(\exists^1(\exists^1\psi(\psi(x, y, z))))(\exists^1\psi(\psi(x, y, z)))\urcorner.$$

This truth value is equal to $(2^8 - 1) - 2^{8-1} = 2^{8-1} - 1$, where $2^8 - 1$ is the designated truth value (verum) at the 8-th level.

It is obvious that if a first-order formula φ has an infinite first-order truth matrix, then we can consider this truth matrix as an infinite 2-adic integer $\alpha \in \mathbf{Z}_2$

In general, vertical quantifiers of the language $\mathscr{L}_{\mathbf{Z}_p}^\infty$ have the following natural *p*-adic valued interpretation:

1. The universal vertical quantifier $\forall_j^{y_j \in \{0, \ldots, p-1\}}(\ldots(\varphi))\ldots(\varphi)$ means that a $(j - 1)$-order formula φ_{j-1} has the truth value $\alpha = (\sum_{k=1}^{j-1} y_k \cdot p^{k-1})$ for any its interpretations. Thus, the universal vertical quantifier $\forall_j^{y_j \in \{0, \ldots, p-1\}}(\ldots(\varphi))\ldots(\varphi)$ means that the satisfiability degree of a first-order formula φ is y_i at any level $i = 1, \ldots, j$.
2. The existential vertical quantifier $\exists_j^{y_j \in \{0, \ldots, p-1\}}(\ldots(\varphi))\ldots(\varphi)$ means that a $(j - 1)$-order formula φ_{j-1} has the truth value $\alpha = (\sum_{k=1}^{j-1} y_k \cdot p^{k-1})$ for some its interpretations and the satisfiability degree of these interpretations for α is y_j.

Definition 8.4 A k-order truth assignment is a function $v_k(\cdot)$ whose domain is the set of all i-order formulas of $\mathscr{L}_{\mathbf{Z}_p}^\infty$ ($i \leq k$) and whose range is the set $\{0, 1, \ldots, p^k - 1\}$ of truth values, such that:

1. $v_1(\varphi)$ is the standard truth assignment of p-valued Łukasiewisz's propositional logic for a proposition φ.

2. For any i-order atomic formula

$$\varphi_i = \forall_{i-1}^{y_{i-1}} (\forall_{i-2}^{y_{i-2}} (\dots (\forall_1^{y_1} \varphi(\varphi))) \dots) \dots (\forall_1^{y_1} \varphi(\varphi)),$$

    a. if $v_{i-1}(\varphi_{i-1}) = y_{i-1} \dots y_1$ for all valuations with a degree $y_i \in \{0, \dots, p - 1\}$, see,[2] then

$$v_i(\varphi_i) = \underbrace{p - 1 \dots p - 1}_{i};$$

    b. if $v_{i-1}(\varphi_{i-1}) = y'_{i-1} \dots y'_1 \neq y_{i-1} \dots y_1$ for *all* valuations with a degree $y'_i \in \{0, \dots, p - 1\}$, then

$$v_i(\varphi_i) = y'_i y'_{i-1} y'_{i-2} \dots y'_1;$$

    c. if $v_{i-1}(\varphi_{i-1}) = y'_{i-1} \dots y'_1 \neq y_{i-1} \dots y_1$ for *some* valuations with a degree $y'_i \in \{0, \dots, p - 1\}$, then

$$v_i(\varphi_i) = y''_i y'_{i-1} y'_{i-2} \dots y'_1,$$

where $y''_i = (p - 1) - y'_i$.

3. For any i-order atomic formula

$$\varphi_i = \exists_{i-1}^{y_{i-1} \in \{0, \dots, p-1\}} (\dots (\exists_1^{y_1 \in \{0, \dots, p-1\}} \varphi(\varphi)) \dots) \dots (\exists_1^{y_1 \in \{0, \dots, p-1\}} \varphi(\varphi)),$$

    a. if $v_{i-1}(\varphi_{i-1}) = y_{i-1} \dots y_1$ for some valuations, then

$$v_i(\varphi_i) = \underbrace{p - 1 \dots p - 1}_{i};$$

    b. if $v_{i-1}(\varphi_{i-1}) = y'_{i-1} \dots y'_1 \neq y_{i-1} \dots y_1$ for *all* valuations with a degree $y_i \in \{0, \dots, p - 1\}$, then

$$v_i(\varphi_i) = y''_i y'_{i-1} y'_{i-2} \dots y'_1,$$

where $y''_i = (p - 1) - y'_i$;

---

[2]In the case of 2-adic valued logic we have only two degrees: 0 and 1. The expression "$v_1(\varphi_1)$ holds for all valuations with the degree 0" means that φ_1 is a contradiction. The expression "$v_1(\varphi_1)$ holds for all valuations with the degree 1" means that φ_1 is a tautology. The expression "$v_{i-1}(\varphi_{i-1})$ holds for all valuations with the degree 0" means that φ_{i-1} is a contradiction. The expression "$v_{i-1}(\varphi_{i-1})$ holds for all valuations with the degree 1" means that φ_{i-1} is a tautology.

c. if $v_{i-1}(\varphi_{i-1}) = y'_{i-1} \ldots y'_1 \neq y_{i-1} \ldots y_1$ for *some* valuations with a degree $y_i \in \{0, \ldots, p-1\}$, then

$$v_i(\varphi_i) = y'_i y'_{i-1} y'_{i-2} \ldots y'_1.$$

4. For any formula φ_i $(i < k)$, $v_k(\varphi_i) = \underbrace{0 \ldots 0}_{k-i} y_{i-1} \ldots y_1 y_0$ iff $v_i(\varphi_i) = y_{i-1} \ldots$ $y_1 y_0$.

5. For any formula φ_i,

$$v_k(\neg \varphi_i) = (p^k - 1) - v_k(\varphi_i).$$

6. For any formulas φ_i and ψ_j such that $\max(i, j) \leq k$,

$$v_k(\varphi_i \supset \psi_j) = ((p^k - 1) - \max(v_k(\varphi_i), v_k(\psi_j)) + v_k(\psi_j)).$$

7. For any formulas φ_i and ψ_j such that $\max(i, j) \leq k$,

$$v_k(\varphi_i \vee \psi_j) = \max(v_k(\varphi_i), v_k(\psi_j)).$$

8. For any formulas φ_i and ψ_j such that $\max(i, j) \leq k$,

$$v_k(\varphi_i \wedge \psi_j) = \min(v_k(\varphi_i), v_k(\psi_j)).$$

Given interpretations of all i-order atomic formulas of $\mathscr{L}_{\mathbf{Z}_p}^\infty$, the corresponding i-order truth assignment would give each i-order atomic formula representing a true statement the value $p^i - 1$, each i-order atomic formula representing a neutral statement the value $\alpha \in \{1, \ldots, p^i - 2\}$, and every i-order atomic formula representing a false statement the value 0.

Also, we obtain the following matrix logic for the language $\mathscr{L}_{\mathbf{Z}_p}^\infty$.

Definition 8.5 The ordered system $(V_i, N_i, I_i, D_i, C_i, \{p^i - 1\})$ is called an i-order *p*-adic valued matrix logic $\mathfrak{M}_{\mathbf{Z}_p}^i$, where

1. $V_i = \{0, \ldots, p^i - 1\} \subset \mathbf{Z}_p$, where i is the finite natural number,
2. $N_i x_i = (p^i - 1) - x_i$ for all $x_i \in V_i$,
   i.e. if $x_i = \alpha_{i-1} \ldots \alpha_2 \alpha_1 \alpha_0$ then

$$N_i x_i = \beta_{i-1} \ldots \beta_2 \beta_1 \beta_0,$$

where $\beta_j = (p-1) - \alpha_j$ for every $j \in \{0, \ldots, i-1\}$,
3. $I_i(x_i, y_i) = ((p^i - 1) - \max(x_i, y_i) + y_i)$ for all $x_i, y_i \in V_i$,
   i.e. if $x_i = \alpha_{i-1} \ldots \alpha_2 \alpha_1 \alpha_0$ and $y_i = \beta_{i-1} \ldots \beta_2 \beta_1 \beta_0$ then

$$I_i(x_i, y_i) = \gamma_{i-1} \ldots \gamma_2 \gamma_1 \gamma_0,$$

where $\gamma_j = ((p-1) - \max(\alpha_j, \beta_j) + \beta_j)$ for every $j \in \{0, \ldots, i-1\}$,

4. $D_i(x_i, y_i) = I_i(I_i(x_i, y_i), y_i) = \max(x_i, y_i)$ for all $x_i, y_i \in V_i$,
   i.e. if $x_i = \alpha_{i-1} \ldots \alpha_2\alpha_1\alpha_0$ and $y_i = \beta_{i-1} \ldots \beta_2\beta_1\beta_0$ then

$$D_i(x_i, y_i) = \gamma_{i-1} \ldots \gamma_2\gamma_1\gamma_0,$$

   where $\gamma_j = \max(\alpha_j, \beta_j)$ for every $j \in \{0, \ldots, i-1\}$,
5. $C_i(x_i, y_i) = N_i D_i(N_i x_i, N_i y_i) = \min(x_i, y_i)$ for all $x_i, y_i \in V_i$,
   i.e. if $x_i = \alpha_{i-1} \ldots \alpha_2\alpha_1\alpha_0$ and $y_i = \beta_{i-1} \ldots \beta_2\beta_1\beta_0$ then

$$C_i(x_i, y_i) = \gamma_{i-1} \ldots \gamma_2\gamma_1\gamma_0,$$

   where $\gamma_j = \min(\alpha_j, \beta_j)$ for every $j \in \{0, \ldots, i-1\}$,
6. $\{p^i - 1\} = \{\alpha_{i-1} \ldots \alpha_2\alpha_1\alpha_0\}$ is the set of designated truth values, where $\alpha_j = p - 1$ for every $j \in \{0, \ldots, i-1\}$.

It can easily be checked that $\mathfrak{M}^i_{\mathbf{Z}_p} = \mathfrak{M}_{\mathbf{Z}_p}$ when $i \to \infty$. Therefore the language $\mathscr{L}^\infty_{\mathbf{Z}_p}$ assumes a p-adic valued logic. Indeed, we can define on the base of $\mathfrak{M}^i_{\mathbf{Z}_p}$ an infinite-order truth assignment that runs over the set of all p-adic integers.

Definition 8.6 An infinite-order truth assignment is a function $v_\infty[\cdot]$ whose domain is the set of all infinite-order formulas of $\mathscr{L}^\infty_{\mathbf{Z}_p}$ and whose range is the set $\{0, \ldots, N_{max}\} = \mathbf{Z}_p$ of truth values, such that:

1. For any i-order formula φ_i, $v_\infty[\varphi_i] = \ldots 0 \ldots 00y_{i-1} \ldots y_1y_0$ iff $v_i(\varphi_i) = y_{i-1} \ldots y_1y_0$.
2. For any infinite-order atomic formula

$$\varphi_\infty = \ldots \forall_{i-1}^{y_{i-1}}(\ldots (\forall_1^{y_1}\varphi(\varphi))\ldots)\ldots(\forall_1^{y_1}\varphi(\varphi)),$$

   a. if $v_\infty[\varphi_\infty] = \ldots y_{i-1} \ldots y_1 \in \mathbf{Z}_p$ for all valuations with a degree $y_i \in \{0, \ldots, p-1\}$, then

$$v_\infty[\varphi_\infty] = \ldots p - 1 \ldots p - 1;$$

   b. if $v_\infty[\varphi_\infty] = \ldots y'_{i-1} \ldots y'_1 \neq \ldots y_{i-1} \ldots y_1$ for *all* valuations with a degree $y'_i \in \{0, \ldots, p-1\}$, then

$$v_\infty[\varphi_\infty] = \ldots y'_i y'_{i-1} y'_{i-2} \ldots y'_1;$$

   c. if $v_\infty[\varphi_\infty] = \ldots y'_{i-1} \ldots y'_1 \neq y_{i-1} \ldots y_1$ for *some* valuations with a degree $y'_i \in \{0, \ldots, p-1\}$, then

$$v_\infty[\varphi_\infty] = \ldots y''_i y'_{i-1} y'_{i-2} \ldots y'_1,$$

   where $y''_i = (p-1) - y'_i$.

3. For any infinite-order atomic formula

$$\varphi_\infty = \dots \exists_{i-1}^{y_{i-1}} (\dots (\exists_1^{y_1} \varphi(\varphi)) \dots) \dots (\exists_1^{y_1} \varphi(\varphi)),$$

a. if $v_\infty[\varphi_\infty] = \dots y_{i-1} \dots y_1 \in \mathbf{Z}_p$ for some valuations with a degree $y_i \in \{0, \dots, p-1\}$, then

$$v_\infty[\varphi_\infty] = \dots p - 1 \dots p - 1;$$

b. if $v_\infty[\varphi_\infty] = \dots y'_{i-1} \dots y'_1 \neq \dots y_{i-1} \dots y_1$ for *all* valuations with a degree $y'_i \in \{0, \dots, p-1\}$, then

$$v_\infty[\varphi_\infty] = \dots y''_i y'_{i-1} y'_{i-2} \dots y'_1,$$

where $y''_i = (p-1) - y'_i$

c. if $v_\infty[\varphi_\infty] = \dots y'_{i-1} \dots y'_1 \neq y_{i-1} \dots y_1$ for *some* valuations with a degree $y'_i \in \{0, \dots, p-1\}$, then

$$v_\infty[\varphi_\infty] = \dots y'_{i-1} y'_{i-1} y'_{i-2} \dots y'_1.$$

4. For any formula φ_∞,
$$v_\infty[\neg\varphi_\infty] = N_{max} - v_\infty[\varphi_\infty].$$

5. For any formulas φ_∞ and ψ_∞,

$$v_\infty[\varphi_\infty \supset \psi_\infty] = (N_{max} - \max(v_\infty[\varphi_\infty], v_\infty[\psi_\infty]) + v_\infty[\psi_\infty]).$$

6. For any formulas φ_∞ and ψ_∞,

$$v_\infty[\varphi_\infty \vee \psi_\infty] = \max(v_\infty[\varphi_\infty], v_\infty[\psi_\infty]).$$

7. For any formulas φ_∞ and ψ_∞,

$$v_\infty[\varphi_\infty \wedge \psi_\infty] = \min(v_\infty[\varphi_\infty], v_\infty[\psi_\infty]).$$

Note the function $v_\infty[\cdot]$ is an infinite sequence of functions $v_i(\cdot)$.

A *p*-adic valued logic that is built on the base of the language $\mathscr{L}_{\mathbf{Z}_p}^\infty$ is denoted by $Ł_{\mathbf{Z}_p}$. Let us introduce some new notations in $Ł_{\mathbf{Z}_p}$:

1. if $v_i(\varphi_i) + v_i(\psi_i) \leq p^i - 1$ (this condition is unnecessary for the infinite-order logic, because the designated truth value is N_{max}), then

$$\varphi_i + \psi_i ::= (\neg\varphi_i \supset \psi_i);$$

2. if $\varphi_i \supset \psi_i \not\equiv \mathbf{p}^i - \mathbf{1}$ (i.e. the implication isn't tautology), then

$$\varphi_i - \psi_i :: = \neg(\varphi_i \supset \psi_i);$$

3. if $\varphi_i \supset \psi_i \not\equiv \mathbf{p}^i - \mathbf{1}, \ldots, (((\varphi_i \supset \psi_i) \supset \ldots) \supset \chi_i) \not\equiv \mathbf{p}^i - \mathbf{1}$, then

$$((\varphi_i - \psi_i) - \ldots) - \chi_i :: = (\neg(\ldots \neg(\varphi_i \supset \psi_i) \supset \ldots) \supset \chi_i);$$

4. if $v_i(\varphi_i) + v_i(\varphi_i) \leq p^i - 1$ (this condition is unnecessary for the infinite-order logic), then

$$\varphi_i + \varphi_i :: = (\neg\varphi_i \supset \varphi_i);$$

5. if $\underbrace{v_i(\varphi_i) + v_i(\varphi_i) + \cdots + v_i(\varphi_i)}_{k} \leq p^i - 1$ (this condition is unnecessary for the

infinite-order logic), then

$$\underbrace{\varphi_i + \varphi_i + \cdots + \varphi_i}_{k} :: = \neg(\ldots \neg(\neg(\neg(\underbrace{\neg\varphi_i \supset \varphi_i) \supset \varphi_i) \supset \varphi_i) \supset \ldots}_{k}) \supset \varphi_i;$$

6. $k \cdot \varphi_i :: = \underbrace{\varphi_i + \varphi_i + \cdots + \varphi_i}_{k}.$

Indeed,

1. in the case $v_i(\varphi_i) + v_i(\psi_i) \leq p^i - 1$ we have $v_i(\psi_i) \leq [(p^i - 1) - v_i(\varphi_i)]$, therefore $(p^i - 1) - \max[(p^i - 1) - v_i(\varphi_i), v_i(\psi_i)] + v_i(\psi_i) = v_i(\varphi_i) + v_i(\psi_i)$.
2. On the other hand, if $v_i(\varphi_i) \not\leq v_i(\psi_i)$, then $\neg(v_i(\varphi_i) \supset v_i(\psi_i)) = (p^i - 1) - [(p^i - 1) - \max(v_i(\varphi_i), v_i(\psi_i)) + v_i(\psi_i)] = v_i(\varphi_i) - v_i(\psi_i)$.
3. Continuing in the same way, we see that notation 3 has also sense.
4. Furthermore, if $v_i(\varphi_i) + v_i(\varphi_i) \leq p^i - 1$, then $v_i(\varphi_i) \leq [(p^i - 1) - v_i(\varphi_i)]$ and $(p^i - 1) - \max[(p^i - 1) - v_i(\varphi_i), v_i(\varphi_i)] + v_i(\varphi_i) = v_i(\varphi_i) + v_i(\varphi_i)$.
5. In the same way we can prove 5.

Notice that the operations $-$, $+$ are the arithmetical subtraction and the arithmetical addition that we define in $Ł_{\mathbf{Z}_p}$ by means of our implication and negation. It is easily shown that all axioms of Peano arithmetic without induction axiom are theorems of infinite-order logic $\mathscr{L}_{\mathbf{Z}_p}^\infty$.

8.3 *p*-Adic Valued Logic of Hilbert's Type

Using the axiomatization of $Ł_p$ created by R. Tuziak (see in [43]), we can set axioms of $Ł_{\mathbf{Z}_p}$. In $Ł_{\mathbf{Z}_p}$ there are four axiom groups:

1. axioms that contain atomic formulas of various order, they are called *mixing axioms*;

2. axioms that contain atomic formulas of the same order, they are called *horizontal axioms*;
3. axioms that contain the same atomic formulas, but they have various order, these axioms are called *vertical*;
4. *axioms of infinite length*.

The first axiom group of calculus $Ł_{\mathbf{Z}_p}$ is as follows:

$$(\varphi_i \supset \phi_j) \supset ((\phi_j \supset \psi_k) \supset (\varphi_i \supset \psi_k)), \tag{8.1}$$

$$\varphi_i \supset (\phi_j \supset \varphi_i), \tag{8.2}$$

$$((\varphi_i \supset \phi_j) \supset \phi_j) \supset ((\phi_j \supset \varphi_i) \supset \varphi_i), \tag{8.3}$$

$$(\varphi_i \wedge \phi_j) \supset \varphi_i, \tag{8.4}$$

$$(\varphi_i \wedge \phi_j) \supset \phi_j, \tag{8.5}$$

$$(\varphi_i \supset \phi_j) \supset ((\varphi_i \supset \psi_k) \supset (\varphi_i \supset (\phi_j \wedge \psi_k))), \tag{8.6}$$

$$\varphi_i \supset (\varphi_i \vee \phi_j), \tag{8.7}$$

$$\phi_j \supset (\varphi_i \vee \phi_j), \tag{8.8}$$

$$(\varphi_i \supset \psi_k) \supset ((\phi_j \supset \psi_k) \supset ((\varphi_i \vee \phi_j) \supset \psi_k)), \tag{8.9}$$

$$(\neg\varphi_i \supset \neg\phi_j) \supset (\phi_j \supset \varphi_i). \tag{8.10}$$

These axioms are called *mixing axioms for* $Ł_{\mathbf{Z}_p}$.

If $i = j = k$ in (8.1)–(8.10), then we obtain *horizontal axioms for* $Ł_{\mathbf{Z}_p}$. The next two horizontal axioms for $Ł_{\mathbf{Z}_p}$ are as follows:

$$(\varphi_i \supset^{p^i} \phi_i) \supset (\varphi_i \supset^{p^i-1} \phi_i), \tag{8.11}$$

here $\varphi_i \supset^0 \phi_i = \phi_i$ and $\varphi_i \supset^{k+1} \phi_i = \varphi_i \supset (\varphi_i \supset^k \phi_i)$,

$$(\varphi_i \equiv (\varphi_i \supset^{s-2} \neg\varphi_i)) \supset^{p^i-1} \varphi_i \tag{8.12}$$

for any $2 \leq s \leq p^i - 1$ such that $p^i - 1$ doesn't divide by s.

In $Ł_{\mathbf{Z}_p}$ there are the following inference rules for finite-order formulas.

1. *Modus ponens*: if two formulas φ_i and $\varphi_i \supset \psi_j$ hold, then we deduce a formula ψ_j:

$$\frac{\varphi_i, \varphi_i \supset \psi_j}{\psi_j}.$$

2. *Substitution rule*:

   a. we can substitute a formula of any order for the atomic formula in mixing axioms;
   b. we can substitute a formula of the same order for the atomic formula in horizontal axioms.

Proposition 8.4 *If $\varphi_j \not\equiv 0$, then we obtain*

$$\vdash_{\text{Ł}_{\mathbf{Z}_p}} \psi_i \supset \underbrace{(\varphi_j + \cdots + \varphi_j)}_{p^i - 1}. \qquad (8.13)$$

Proof Introduce the following notation:

$$\mathbf{p^i} - \mathbf{1}{::} = \varphi_i \supset (\phi_i \supset \varphi_i)$$

Then our formula (8.13) can be understood as follows:

$$\psi_i \supset \underbrace{(\mathbf{p^i} - \mathbf{1} + \cdots + \mathbf{p^i} - \mathbf{1})}_{\varphi_j}$$

It has the following complete notation:

$$\psi_i \supset (\neg(\ldots \neg(\neg(\neg(\underbrace{\neg \mathbf{p^i} - \mathbf{1} \supset \mathbf{p^i} - \mathbf{1}) \supset \mathbf{p^i} - \mathbf{1}) \supset \mathbf{p^i} - \mathbf{1}) \supset \ldots) \supset \mathbf{p^i} - \mathbf{1})}_{\varphi_j}$$

$$(8.14)$$

Consider the *i*-order theorem:

$$\psi_i \supset (\chi_i \supset (\varphi_i \supset (\phi_i \supset \varphi_i))).$$

If we substitute the expression

$$\neg(\ldots \neg(\neg(\neg(\underbrace{\neg \mathbf{p^i} - \mathbf{1} \supset \mathbf{p^i} - \mathbf{1}) \supset \mathbf{p^i} - \mathbf{1}) \supset \mathbf{p^i} - \mathbf{1}) \supset \ldots)}_{v_j(\varphi_j) - 1}$$

for the atom χ_i, then we obtain (8.14).   □

Now we propose *vertical axioms* on the basis of which we shall put all sequences of the form ψ_∞.

Consider the following formula

$$\psi_{i+1} \equiv \psi_i + \underbrace{\varphi_1 + \cdots + \varphi_1}_{p^i}.$$

Let a_i, a_{i+1}, b_1 be matrix values of ψ_i, ψ_{i+1}, φ_1 respectively. Then we deduce the following equality:

$$a_{i+1} = a_i + p^i \cdot b_1$$

It can easily be checked that b_1 runs from 0 to $p - 1$ and a_{i+1} runs over all elements of p sets:

1. $p^{i+1} - p^i \leqslant a_{i+1} < p^{i+1}$;

...

$p. \ 0 \leqslant a_{i+1} < \underbrace{(\ldots (p^{i+1} - p^i) - \ldots) - p^i}_{p-1}.$

We see that there exists a unique solution a_i in accordance with a particular value of the truth variable a_{i+1} and a unique solution a_{i+1} in accordance with a particular value of the truth variable a_i.

Thus, we obtain the next axiom:

$$(((\mathbf{p}^{i+1} - \mathbf{1}) - (\mathbf{p}^i - \mathbf{1})) \supset \psi_{i+1}) \supset (\psi_{i+1} \equiv$$
$$\equiv \underbrace{((\mathbf{p} - \mathbf{1}) + \cdots + (\mathbf{p} - \mathbf{1}))}_{p^i} + \psi_i), \tag{8.15}$$

where $\mathbf{p} - \mathbf{1}$ is a tautology at the first-order level and $\mathbf{p}^i - \mathbf{1}$ (respectively $\mathbf{p}^{i+1} - \mathbf{1}$) is a tautology at i-th order level (respectively at $(i + 1)$-th order level).

Further we have the following axioms:

$$(\psi_{i+1} \equiv \underbrace{(\mathbf{k} + \cdots + \mathbf{k})}_{p^i} + \psi_i) \supset$$
$$\supset ((((\ldots ((\mathbf{p}^{i+1} - \mathbf{1}) - \underbrace{(\mathbf{p}^i - \mathbf{1})) - \ldots) - (\mathbf{p}^i - \mathbf{1}))}_{0 < p-k \leq p} - \neg\mathbf{k}) \supset \psi_{i+1}), \tag{8.16}$$

$$(\psi_{i+1} \equiv \underbrace{(\mathbf{k} + \cdots + \mathbf{k})}_{p^i} + \psi_i) \supset$$
$$\supset (\psi_{i+1} \supset (((\ldots ((\mathbf{p}^{i+1} - \mathbf{1}) - \underbrace{(\mathbf{p}^i - \mathbf{1})) - \ldots) - (\mathbf{p}^i - \mathbf{1}))}_{0 \leq (p-1)-k \leq p-1} - \neg\mathbf{k})), \tag{8.17}$$

where $\neg\mathbf{k}$ is a first-order formula that has the truth value $((p - 1) - k) \in \{0, \ldots, p - 1\}$ for any its interpretations and $\mathbf{k}$ is a first-order formula that has the truth value $k \in \{0, \ldots, p - 1\}$ for any its interpretations.

$$(\psi_{i+1} \supset (\mathbf{p}^i - \mathbf{1})) \supset (\psi_{i+1} \equiv \psi_i) \tag{8.18}$$

$$\underbrace{\psi_{i+1} \equiv \psi_i \vee \psi_{i+1} \equiv (\psi_i + p^i \cdot \mathbf{1}) \vee \ldots \vee \psi_{i+1} \equiv (\psi_i + p^i \cdot (\mathbf{p} - \mathbf{1}))}_{p}, \tag{8.19}$$

where $\mathbf{1}$ is a first-order formula that has the truth value 1 for any its interpretations, $\mathbf{p} - \mathbf{1}$ is a first-order formula that has the truth value $p - 1$ for any its interpretations, etc.

The axioms (8.15)–(8.19) are called *vertical*. We use only modus ponens for vertical axioms.

Now we get the following notation:

$$\psi_\infty :: = \psi_1 + \underbrace{\beta_1 + \cdots + \beta_1}_{p} + \underbrace{\gamma_1 + \cdots + \gamma_1}_{p^2} + \cdots + \underbrace{\delta_1 + \cdots + \delta_1}_{p^i} + \ldots ,(8.20)$$

where

1. $\psi_1, \beta_1, \gamma_1, \delta_1$ are some first-order formulas that we obtain by means of the axioms (8.15)–(8.19),

2. $\psi_1 + \underbrace{\beta_1 + \cdots + \beta_1}_{p} \equiv \psi_2,$

3. $\psi_1 + \underbrace{\beta_1 + \cdots + \beta_1}_{p} + \underbrace{\gamma_1 + \cdots + \gamma_1}_{p^2} \equiv \psi_3,$

4. $\psi_1 + \underbrace{\beta_1 + \cdots + \beta_1}_{p} + \underbrace{\gamma_1 + \cdots + \gamma_1}_{p^2} + \cdots + \underbrace{\delta_1 + \cdots + \delta_1}_{p^i} \equiv \psi_{i+1}$, etc.

Proposition 8.5 $\vdash_{Ł_{\mathbf{Z}_p}} \varphi_i \supset \varphi_\infty.$

Proof Trivially, it follows direct from the proposition of infinite-order language

$$\varphi_i \supset (\varphi_i + \psi_\infty)$$

that is theorem in $Ł_{\mathbf{Z}_p}$.   □

The *axioms of infinite length* are as follows:

$$(\varphi_\infty \supset \phi_\infty) \supset ((\phi_\infty \supset \psi_\infty) \supset (\varphi_\infty \supset \psi_\infty)), \tag{8.21}$$

$$\varphi_\infty \supset (\phi_\infty \supset \varphi_\infty), \tag{8.22}$$

$$((\varphi_\infty \supset \phi_\infty) \supset \phi_\infty) \supset ((\phi_\infty \supset \varphi_\infty) \supset \varphi_\infty), \tag{8.23}$$

$$(\varphi_\infty \wedge \phi_\infty) \supset \varphi_\infty, \tag{8.24}$$

$$(\varphi_\infty \wedge \phi_\infty) \supset \phi_\infty, \tag{8.25}$$

$$(\varphi_\infty \supset \phi_\infty) \supset ((\varphi_\infty \supset \psi_\infty) \supset (\varphi_\infty \supset (\phi_\infty \wedge \psi_\infty))), \tag{8.26}$$

$$\varphi_\infty \supset (\varphi_\infty \vee \phi_\infty), \tag{8.27}$$

$$\phi_\infty \supset (\varphi_\infty \vee \phi_\infty), \tag{8.28}$$

$$(\varphi_\infty \supset \psi_\infty) \supset ((\phi_\infty \supset \psi_\infty) \supset ((\varphi_\infty \vee \phi_\infty) \supset \psi_\infty)), \tag{8.29}$$

$$(\neg\varphi_\infty \supset \neg\phi_\infty) \supset (\phi_\infty \supset \varphi_\infty), \tag{8.30}$$

In $Ł_{Z_p}$ there are the following inference rules for infinite-order formulas.

1. *Modus ponens*: if two formulas φ_∞ and $\varphi_\infty \supset \psi_\infty$ hold, then we deduce a formula ψ_∞:

$$\frac{\varphi_\infty, \varphi_\infty \supset \psi_\infty}{\psi_\infty}.$$

2. *Substitution rule*: we can substitute every infinite-order formula for the infinite-order atomic formula.

Theorem 1 (Soundness theorem) *If a formula φ is an axiom (or a theorem) in $Ł_{Z_p}$, then φ is a tautology in $\mathfrak{M}_{Z_p}$, i.e. from $\vdash_{Ł_{Z_p}} \varphi$ it follows that $\models_{\mathfrak{M}_{Z_p}} \varphi$.*

Proof This theorem can be proved by constructing of *p*-adic truth tables for the expressions of (8.1)–(8.12), (8.15)–(8.19), (8.21)–(8.30). If the formulas of (8.1)–(8.12), (8.15)–(8.19) have *i*-th order (i.e. if *i* is the maximal order of their atoms), then the maximal truth value for atoms in (8.1)–(8.12), (8.15)–(8.19) is $p^i - 1$. The maximal truth value for atoms in (8.21)–(8.30) is N_{max}. At the same time, inference rules preserve tautologies. □

8.4 *p*-Adic Probability Theory

Consider some principal definitions of *p*-adic measure theory.

Let X be an arbitrary set and let $\mathscr{R}$ be a ring of subsets of X, i.e. union, intersection, and difference of two sets of X also belong to $\mathscr{R}$. The pair $(X, \mathscr{R})$ is called a *measurable space*.

The ring $\mathscr{R}$ is said to be *separating* if, for every two distinct elements x, y of X, there exists a set $A \in \mathscr{R}$ such that $x \in A$, $y \notin A$.

Definition 8.7 Let us consider a measurable space $(X, \mathscr{R})$. Suppose that $\mathscr{R}$ is a separating covering ring of a set X. Then a measure on a ring $\mathscr{R}$ is a map $\mu: \mathscr{R} \to \mathbf{Q}_p$ with the properties:

1. μ is additive;
2. for all $A \in \mathscr{R}$, $\sup\{|\mu(B)|_p : B \in \mathscr{R}, B \subset A\} < \infty$;

   Notice that

1. using condition 1 of this definition, we get
   $\mu(A \cup B) = \mu(A) + \mu(B)$ for all $A, B \in \mathscr{R}$, $A \cap B = \emptyset$
2. under condition 2, for all $n \in \mathbf{N} \cup \{0\}$ there exists a number (constant) C such that $|\mu(p^n \cdot \mathbf{Z}_p)|_p < C$.   Otherwise, $p^n \cdot \mathbf{Z}_p = \{x \in \mathbf{Q}_p: \rho_p(x, 0) \leq p^{-n}\} = B[0, p^{-n}]$. Thus, for all $n \in \mathbf{N} \cup \{0\}$ there exists a number (constant) C such that $|\mu(B[0, p^{-n}])|_p < C$.

Suppose that all elements of the ring $\mathscr{R}$ are μ-measurable sets. By $\mathscr{R}_\mu$ denote this ring. A ring $\mathscr{R}$ is called an algebra $\mathscr{A}$ of subsets of X if $\emptyset$, X belong to $\mathscr{R}$.

Definition 8.8 Let $\mu: \mathscr{A} \to \mathbf{Q}_p$ be a measure defined on a separating algebra $\mathscr{A}$ of subsets of the set X which satisfies the normalization condition $\mu(X) = 1$. We set $\mathscr{F} = \mathscr{A}_\mu$ and denote the extension of μ on $\mathscr{F}$ by the symbol $\mathbf{P}$. A triple $(X, \mathscr{F}, \mathbf{P})$ is said to be a p-adic *probability space*, where X is a sample space, $\mathscr{F}$ is an algebra of events, $\mathbf{P}$ is a probability.

This definition was given by Khrennikov [48].

Notice that philosophical and logical fundamentals of p-adic probability theory were discussed in [49].

Transform the matrix logic $\mathfrak{M}_{\mathbf{Z}_p}$ into a p-adic probability theory. Let us remember that a formula $\varphi \in \mathbf{L}_{\mathbf{Z}_p}$ has the truth value $0 \in \mathbf{Z}_p$ in $\mathfrak{M}_{\mathbf{Z}_p}$ if φ is false, a formula $\varphi \in \mathbf{L}_{\mathbf{Z}_p}$ has the truth value $N_{max} \in \mathbf{Z}_p$ in $\mathfrak{M}_{\mathbf{Z}_p}$ if φ is true, and a formula $\varphi \in \mathbf{L}_{\mathbf{Z}_p}$ has other truth values $\alpha \in \mathbf{Z}_p$ in $\mathfrak{M}_{\mathbf{Z}_p}$ if φ is neutral.

Definition 8.9 A function $\mathbf{P}(\varphi)$ is said to be a probability measure of a formula φ in $\mathfrak{M}_{\mathbf{Z}_p}$ if $\mathbf{P}(\varphi)$ ranges over numbers of $\mathbf{Q}_p$ and satisfies the following axioms:

1. $\mathbf{P}(\varphi) = \frac{\alpha}{N_{max}}$, where α is a truth value of φ;
2. if a conjunction $\varphi \wedge \psi$ has the truth value 0, then $\mathbf{P}(\varphi \vee \psi) = \mathbf{P}(\varphi) + \mathbf{P}(\psi)$,
3. $\mathbf{P}(\varphi \wedge \psi) = \min(\mathbf{P}(\varphi), \mathbf{P}(\psi))$.

Notice that

1. taking into account condition 1 of our definition, if φ has the truth value N_{max} for any its interpretations, i.e. φ is a tautology, then $\mathbf{P}(\varphi) = 1$, and if φ has the truth value 0 for any its interpretations, i.e. φ is a contradiction, then $\mathbf{P}(\varphi) = 0$;
2. since $\mathbf{P}(\varphi) + \mathbf{P}(\neg\varphi) = 1$, we obtain $\mathbf{P}(\neg\varphi) = 1 - \mathbf{P}(\varphi)$.

From $\mathbf{P}(N_{max}) = 1$ it follows that

$$\mathbf{P}(\max\{x \in V_{\mathbf{Z}_p}\}) = \sum_{x \in V_{\mathbf{Z}_p}} \mathbf{P}(x) = 1$$

All events have a conditional plausibility in the logical theory of p-adic probability:

$$\mathbf{P}(\varphi) \equiv \mathbf{P}(\varphi/N_{max}),$$

i.e. for any φ, we consider the conditional plausibility that there is an event of φ, given an event N_{max},

$$\mathbf{P}(\varphi/\psi) = \frac{\mathbf{P}(\varphi \wedge \psi)}{\mathbf{P}(\psi)}.$$

In the next chapter, we are going to show how p-adic arithmetic operations defined in logic $Ł_{\mathbb{Z}_p}$ can be implemented in *Physarum* machines. In Chap. 10, we will consider how p-adic valued probabilities can be used in defining game strategies for the slime mould.

Chapter 9
p-Adic Valued Arithmetic Gates

9.1 *p*-Adic Valued Physarum Machines

Let $\mathcal{PM} = (Ph, Attr, Rep)$ be a structure of the *Physarum* machine. In Sect. 3, we have described a dynamics of the *Physarum* machine $\mathcal{PM}$ by the family of the sets of protoplasmic veins formed by plasmodium during its action. In this chapter, the dynamics of the *Physarum* machine $\mathcal{PM}$ will be additionally considered by means of two sets of attractants (occupied by plasmodium and unoccupied) determined for each time instant t. Therefore, the set *Attr* of attractants, at each time instant t, can be divided into two disjoint sets:

- $Attr_{\bullet}^{t}$—the set of attractants occupied by the plasmodium at t,
- $Attr_{o}^{t}$—the set of unoccupied attractants at t.

The cardinality of $Attr_{\bullet}^{t}$ will be marked with $m_{\bullet}^{t}$ whereas the cardinality of $Attr_{o}^{t}$ will be marked with m_{o}^{t}. It is easy to see that, for each time instant t:

- $Attr_{\bullet}^{t} \cup Attr_{o}^{t} = Attr$,
- $Attr_{\bullet}^{t} \cap Attr_{o}^{t} = \emptyset$,
- $m_{\bullet}^{t} + m_{o}^{t} = m$.

The set Π^{t} of all active points at time instant t consists of the set *Ph* of original points of the plasmodium as well as the set $Attr_{\bullet}^{t}$ of all attractants occupied by the plasmodium at t. Hence, the cardinality of Π^{t} is equal to $k + m_{\bullet}^{t}$, where $k = card(Ph)$.

Let $\{\Pi^{t}\}_{t=0,1,2,\dots}$ be a family of the sets of all active points at time instants $t = 0, 1, 2, \dots$ in the *Physarum* machine $\mathcal{PM}$. A dynamics of $\mathcal{PM}$ over time is defined by the family $V = \{V^{t}\}_{t=0,1,2,\dots}$ of the sets of protoplasmic veins propagated by the plasmodium, where $V^{t} = \{v_{1}^{t}, v_{2}^{t}, \dots, v_{r_{t}}^{t}\}$ is the set of all protoplasmic veins of the plasmodium present at time instant t in $\mathcal{PM}$. Each vein $v_{i}^{t} \in V^{t}$, where $i = 1, 2, \dots, r_{t}$, is an unordered pair $\{\pi_{j}^{t}, \pi_{k}^{t}\}$ of two adjacent active points $\pi_{j}^{t} \in \Pi^{t}$ and $\pi_{k}^{t} \in \Pi^{t}$, connected directly by v_{i}^{t}.

© Springer International Publishing AG, part of Springer Nature 2019
A. Schumann and K. Pancerz, *High-Level Models of Unconventional
Computations*, Studies in Systems, Decision and Control 159,
https://doi.org/10.1007/978-3-319-91773-3_9

Fig. 9.1   An elementary
block

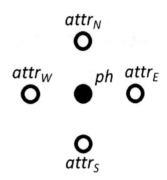

In the context of the dynamics of the *Physarum* machine, we can determine a simple growing path that can be observed at a given time instant t.

Definition 9.1   Let $\{\Pi^t\}_{t=0,1,2,\ldots}$ be a family of the sets of all active points at time instants $t = 0, 1, 2, \ldots$ in the *Physarum* machine $\mathscr{PM}$. A simple growing path of length $l - 1$ is a set of veins $\langle\{\pi_{i_1}^t, \pi_{i_2}^t\}, \{\pi_{i_2}^t, \pi_{i_3}^t\}, \ldots, \{\pi_{i_{l-1}}^t, \pi_{i_l}^t\}\rangle$, at time instant t, where $\pi_{i_1}^t, \pi_{i_2}^t, \ldots \pi_{i_l}^t \in \Pi^t$, $\pi_{i_1}^t$ is an original point of the plasmodium, $\pi_{i_1}^t \neq \pi_{i_l}^t$, and all veins are pairwise different.

The environment of slime mould can contain many attractants. In this case, the plasmodium can simultaneously propagate its networks in many different directions. Let us assume that the plasmodium, at time instant t, in an active point $\pi_i^t \in \Pi^t$, where $i = 1, 2, \ldots, k + m_{\bullet}^t$, cannot see more than $p - 1$ attractants. Therefore, we can build up the finite part of p-adic arithmetic. The *Physarum* machines are called *p-adic* if, at each active point from $\{\Pi^t\}_{t=0,1,2,\ldots}$ and at each time instant t, the plasmodium can see not more than $p - 1$ attractants.

Without loss of generality, let us consider an elementary block shown in Fig. 9.1, i.e. the block containing just 4 attractants. So, we can focus on the case of finite 5-adic arithmetic [125]. Formally, a structure of the elementary block of the *Physarum* machine can be described as a triple $\mathscr{PM}_{el} = (Ph_{el}, Attr_{el}, Rep_{el})$, where:

- $Ph_{el} = \{ph\}$,
- $Attr_{el} = \{attr_N, attr_E, attr_S, attr_W\}$,
- $Rep_{el} = \emptyset$.

It is easy to see that, in case of the considered elementary block $\mathscr{PM}_{el}$, plasmodium can move from the origin point ph to the north (i.e. to the attractant $attr_N$), to the east (i.e. to the attractant $attr_E$), to the south (i.e. to the attractant $attr_S$), and to the west (i.e. to the attractant $attr_W$). Examples of simple growing paths of length 1, which can be propagated by the plasmodium in $\mathscr{PM}_{el}$, are shown in Fig. 9.2).

In Fig. 9.3, we have shown an illustrative example of dynamics of the elementary block $\mathscr{PM}_{el}$.

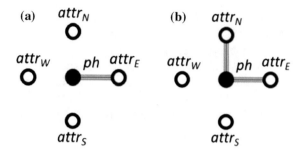

Fig. 9.2 Examples of simple growing paths of length 1 in the elementary block $\mathscr{P}\mathscr{M}_{el}$: **a** one path, **b** two paths

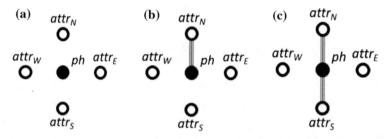

Fig. 9.3 An illustrative example of dynamics of the elementary block $\mathscr{P}\mathscr{M}_{el}$: **a** $\mathscr{P}\mathscr{M}_{el}$ observed at time instant t_0, **b** $\mathscr{P}\mathscr{M}_{el}$ observed at time instant t_1, **c** $\mathscr{P}\mathscr{M}_{el}$ machine observed at time instant t_2

The sets of attractants, at each time instant, are as follows:

- $Attr_{\bullet}^0 = \emptyset$, $Attr_{\circ}^0 = \{attr_N, attr_E, attr_S, attr_W\}$,
- $Attr_{\bullet}^1 = \{attr_N\}$, $Attr_{\circ}^1 = \{attr_E, attr_S, attr_W\}$,
- $Attr_{\bullet}^2 = \{attr_N, attr_S\}$, $Attr_{\circ}^2 = \{attr_E, attr_S\}$.

Hence, $m_{\bullet}^0 = 0$, $m_{\circ}^0 = 1$, $m_{\bullet}^1 = 1$, $m_{\circ}^1 = 3$, $m_{\bullet}^2 = 2$, and $m_{\circ}^2 = 2$. Hence, we obtain the following sets of all active points:

- $\Pi_{\bullet}^0 = \{ph\}$,
- $\Pi_{\bullet}^1 = \{ph, attr_N\}$,
- $\Pi_{\bullet}^2 = \{ph, attr_N, attr_S\}$.

The cardinality of $\Pi_{\bullet}^0$ is equal to 1, the cardinality of $\Pi_{\bullet}^1$ is equal to 2, and the cardinality of $\Pi_{\circ}^2$ is equal to 3.

Finally, for our elementary block $\mathscr{P}\mathscr{M}_{el}$, we have $V = \{V^t\}_{t=0,1,2}$, where:

- $V^0 = \emptyset$,
- $V^1 = \{\{ph, attr_N\}\}$,
- $V^2 = \{\{ph, attr_N\}, \{ph, attr_S\}\}$.

We can estimate maximal numbers of all simple growing paths that can be created in the elementary block $\mathscr{PM}_{el}$. Therefore, we can code information about numbers of simple growing paths in terms of *p*-adic strings.

In case of the elementary block $\mathscr{PM}_{el}$, we are limited by 5-adic one bit strings $\alpha^t = \alpha_0^t$, where α_0^t is a number of simple growing paths of length 1 at time instant t, because the set of all possible simple growing paths of length 1, that can be created in $\mathscr{PM}_{el}$, includes $\langle\{ph, attr_N\}\rangle$, $\langle\{ph, attr_E\}\rangle$, $\langle\{ph, attr_S\}\rangle$, and $\langle\{ph, attr_W\}\rangle$. One can see that:

- $\alpha^0 = 0$,
- $\alpha^1 = 1$,
- $\alpha^2 = 2$.

9.2 *p*-Adic Valued Adder and Subtracter

The elementary block $\mathscr{PM}_{el}$, described in Sect. 9.1, is a building block for *p*-adic valued arithmetic gates. In the presented approach, *p*-adic valued arithmetic gates are created as vector or matrix structures, i.e. they are composed of single rows of elemntary blocks or rectangular grids of elementary blocks.

The addition can be caused by fusion of the plasmodium. A *p*-adic valued adder, where $p = 5$, can be created as a two block vector adder $\mathscr{PM}_{2bva}$ shown in Fig. 9.4. In this adder, two simple growing paths are fused in the attractant $attr_{E_1}/attr_{W_2}$. If two simple growing paths meet this attractant, then we treat this situation as the 5-adic valued addition.

It is easy to estimate maximal numbers of all simple growing paths that can be propagated in the two block vector adder $\mathscr{PM}_{2bva}$. The possible simple growing paths starting at the original point ph_1 are as follows:

- simple growing paths of length 1:

  - $\langle\{ph_1, attr_{N_1}\}\rangle$,

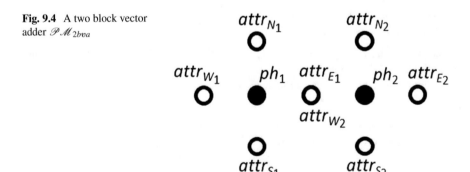

Fig. 9.4 A two block vector adder $\mathscr{PM}_{2bva}$

- $\langle\{ph_1, attr_{E_1}\}\rangle$,
- $\langle\{ph_1, attr_{S_1}\}\rangle$,
- and $\langle\{ph_1, attr_{W_1}\}\rangle$.

- simple growing paths of length 2:

  - only $\langle\{ph_1, attr_{E_1}\}, \{attr_{E_1}, ph_2\}\rangle$,

- simple growing paths of length 3:

  - $\langle\{ph_1, attr_{E_1}\}, \{attr_{E_1}, ph_2\}, \{ph_2, attr_{N_2}\}\rangle$,
  - $\langle\{ph_1, attr_{E_1}\}, \{attr_{E_1}, ph_2\}, \{ph_2, attr_{E_2}\}\rangle$,
  - and $\langle\{ph_1, attr_{E_1}\}, \{attr_{E_1}, ph_2\}, \{ph_2, attr_{S_2}\}\rangle$.

Analogously, the possible simple growing paths starting at the original point ph_2 are as follows:

- simple growing paths of length 1:

  - $\langle\{ph_2, attr_{N_2}\}\rangle$,
  - $\langle\{ph_2, attr_{E_2}\}\rangle$,
  - $\langle\{ph_2, attr_{S_2}\}\rangle$,
  - and $\langle\{ph_2, attr_{W_2}\}\rangle$.

- simple growing paths of length 2:

  - only $\langle\{ph_2, attr_{W_2}\}, \{attr_{W_2}, ph_1\}\rangle$,

- simple growing paths of length 3:

  - $\langle\{ph_2, attr_{W_2}\}, \{attr_{W_2}, ph_1\}, \{ph_1, attr_{N_1}\}\rangle$,
  - $\langle\{ph_2, attr_{W_2}\}, \{attr_{W_2}, ph_1\}, \{ph_1, attr_{W_1}\}\rangle$,
  - and $\langle\{ph_2, attr_{W_2}\}, \{attr_{W_2}, ph_1\}, \{ph_1, attr_{S_1}\}\rangle$.

Therefore, we are limited by the 5-adic integer $4 + 4 = 13$. This integer is obtained by the adder if and only if we have 8 paths of length 1. The two block vector adder $\mathscr{P}\mathscr{M}_{2bva}$ gives a 5-adic integer $x + y$ at time t if and only if we have $x + y$ paths of length 1 at the same time t.

Sometimes, we need information about a configuration of plasmodium veins. In this case, we can code adders by larger *p*-adic integers $\ldots \alpha_2^t \, \alpha_1^t \, \alpha_0^t$, where α_i^t is a number of paths of lengths $i + 1$. For example, for the adder $\mathscr{P}\mathscr{M}_{2bva}$ we obtain 9-adic three bit strings $\alpha^t = \alpha_2^t \, \alpha_1^t \, \alpha_0^t$, where:

- α_2^t is a number of simple paths of length 3 at time instant t,
- α_1^t is a number of simple paths of length 2 at time instant t,
- α_0^t is a number of simple paths of length 1 at time instant t.

An illustrative example of dynamics of the two block vector adder $\mathscr{P}\mathscr{M}_{2bva}$ is shown in Fig. 9.5. The example explains how the addition is done.

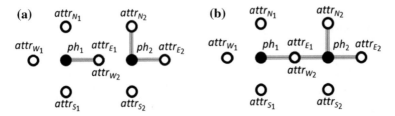

Fig. 9.5 An illustrative example of dynamics of the two block vector adder $\mathscr{PM}_{2bva}$: **a** a situation before the addition, at time instant 0, **b** a situation after the addition, at time instant 1. The addition holds if and only if $attr_{W_2}/attr_{E_1}$ is occupied

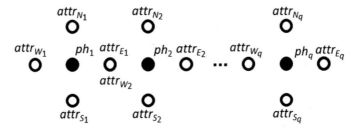

Fig. 9.6 The q block vector adder

Before the addition, we obtain the following strings:

- for the first elementary block: $\alpha^0 = 1$,
- for the second elementary block: $\alpha^0 = 2$.

After the addition, we obtain the 5-adic integer $3 + 1 = 4$. An appropriate adder can be coded by the 9-adic string $\alpha^1 = 2\ 2\ 4$, because:

- there are four simple growing paths of length 1:

  - $\langle\{ph_1, attr_{E_1}\}\rangle$,
  - $\langle\{ph_2, attr_{N_2}\}\rangle$,
  - $\langle\{ph_2, attr_{E_2}\}\rangle$,
  - and $\langle\{ph_2, attr_{W_2}\}\rangle$.

- there are two simple growing paths of length 2:

  - $\langle\{ph_1, attr_{E_1}\}, \{attr_{E_1}, ph_2\}\rangle$,
  - and $\langle\{ph_2, attr_{W_2}\}, \{attr_{W_2}, ph_1\}\rangle$.

- there are two simple growing paths of length 3:

  - $\langle\{ph_1, attr_{E_1}\}, \{attr_{E_1}, ph_2\}, \{ph_2, attr_{N_2}\}\rangle$,
  - and $\langle\{ph_1, attr_{E_1}\}, \{attr_{E_1}, ph_2\}, \{ph_2, attr_{E_2}\}\rangle$.

Let us consider a more general case of a block vector adder shown in Fig. 9.6.

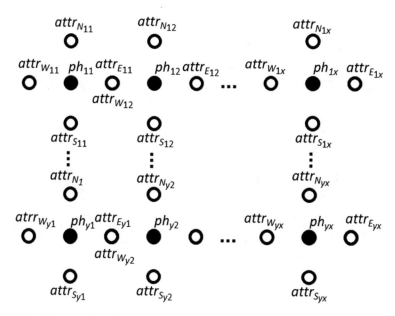

Fig. 9.7 A *q* block matrix adder

In case of the *q* block vector adder, we are limited by the 5-adic integer $\underbrace{4 + \cdots + 4}_{q}$

and this adder is coded by $4q + 1$-adic $2q - 1$ bit strings $\alpha^t = \alpha^t_{2q-2} \cdots \alpha^t_3 \alpha^t_2 \alpha^t_1 \alpha^t_0$, where:

- α^t_{2q-2} is a number of simple growing paths of length $2q - 1$ at time instant t,
- ...,
- α^t_3 is a number of simple growing paths of length 4 at time instant t,
- α^t_2 is a number of simple growing paths of length 3 at time instant t,
- α^t_1 is a number of simple growing paths of length 2 at time instant t,
- α^t_0 is a number of simple growing paths of length 1 at time instant t.

As previously, it is easy to estimate the maximal numbers of all simple growing paths that can be propagated in the *q* block vector adder. The upper bound is determined by the number of all possible simple growing paths of length 1 that can be propagated by the plasmodium. This number is equal to $4q$.

The generalized *q* block matrix adder with *x* columns and *y* rows is shown in Fig. 9.7.

The upper bound of numbers of all simple growing paths is determined by the number of all possible simple growing paths of length 1 that can be propagated by the plasmodium. This number is equal to $4xy$. Therefore, in case of the *q* block matrix adder, we are limited by the 5-adic integer $\underbrace{4 + \cdots + 4}_{xy}$ and this adder is coded

by $4xy + 1$-adic $2(x - 1) + 2(y - 1) + 1$ bit strings. A number of bits comes from

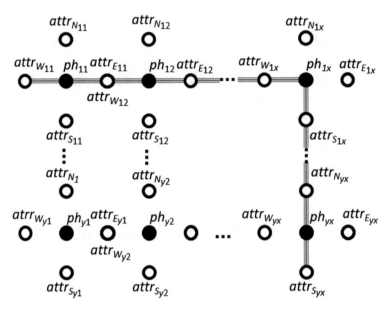

Fig. 9.8 One of the possibly longest simple growing paths that can be propagated in q block matrix adder

Fig. 9.9 A two block vector subtracter

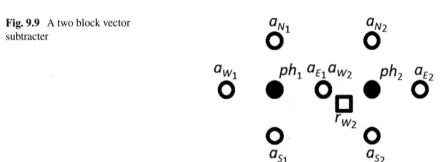

the fact that the longest simple growing paths can be propagated between top left and bottom right elementary blocks or between top right and bottom left elementary blocks. One of the possibly longest simple growing paths is shown in Fig. 9.8.

The subtraction can be caused by the repelling of the plasmodium. A p-adic valued subtracter, where $p = 5$, can be created as a two block vector subtracter $\mathscr{P}\mathscr{M}_{2bvs}$ shown in Fig. 9.9. In this subtracter, the repellent rep_{W_2} is placed to make the subtraction. Activation of this repellent causes that the protoplasmic vein between ph_2 and $attr_{W_2}$ is annihilated.

An illustrative example of dynamics of the two block vector subtracter $\mathscr{P}\mathscr{M}_{2bvs}$ is shown in Fig. 9.10. The example explains how the subtraction is done.

Before the subtraction (i.e. when the repellent rep_{W_2} is inactive), we have the 5-adic integer 4 obtained by the adder coded by the string $\alpha^0 = 2\ 2\ 4$. After the

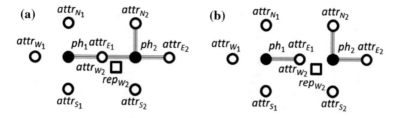

Fig. 9.10 An illustrative example of dynamics of the two block vector subtracter $\mathscr{PM}_{2bvs}$: **a** a situation before the subtraction, at time instant 0, **b** a situation after the subtraction, at time instant 1

subtraction (i.e. when the repellent rep_{W_2} is active), we obtain the 5-adic integer 3 coded by the following strings:

- for the first elementary block: $\alpha^1 = 1$,
- for the second elementary block: $\alpha^1 = 2$.

9.3 High-Level Models of *p*-adic Valued Arithmetic Gates

In this section, we present high-level models (in the form of timed transition systems and Petri nets) of *p*-adic valued arithmetic gates described in Sect. 9.2. Such models can be used to program *Physarum* machines (see Sect. 7).

In Figs. 9.11 and 9.12, timed transition system models of a two block vector adder and a two block vector subtracter, respectively, are shown. Formally, we have:

- a two block vector adder: $TS_a = (S_a, E_a, T_a, I_a)$, where

  - $S_a = \{s_{ph1}, s_{ph2}, s_{aS1}, s_{aW1}, s_{aN1}, s_{aE2}, s_{aS2}, s_{aN2}, s_{aF12}\}$,
  - $E_a = \{e_{E1}, e_{S1}, e_{W1}, e_{N1}, e_{E2}, e_{S2}, e_{W2}, e_{N2}\}$,

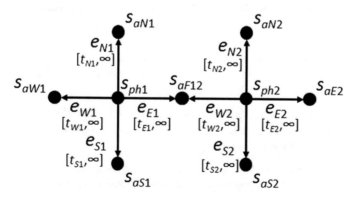

Fig. 9.11 A timed transition system model of a two block vector adder

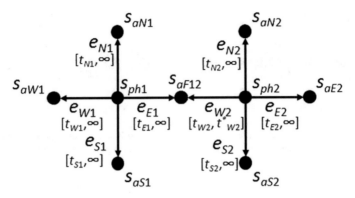

Fig. 9.12 A timed transition system model of a two block vector subtracter

- $T_a = \{(s_{ph1}, e_{E1}, s_{aF12}),\ (s_{ph1}, e_{S1}, s_{aS1}),\ (s_{ph1}, e_{W1}, s_{aW1}),\ (s_{ph1}, e_{N1}, s_{aN1}),$
  $(s_{ph2}, e_{E2}, s_{aE2}), (s_{ph2}, e_{S2}, s_{aS2}), (s_{ph2}, e_{W2}, s_{aF12}), (s_{ph2}, e_{N2}, s_{aN2})\}$,
- $I_a = \{s_{ph1}, s_{ph2}\}$,
- and $l(e_{E1}) = t_{E1},\ \ l(e_{S1}) = t_{S1},\ \ l(e_{W1}) = t_{W1},\ \ l(e_{N1}) = t_{N1},\ \ l(e_{E2}) = t_{E2},$
  $l(e_{S2}) = t_{S2}, l(e_{W2}) = t_{W2}, l(e_{N2}) = t_{N2}, u(e_{E1}) = \infty, u(e_{S1}) = \infty, u(e_{W1}) =$
  $\infty, u(e_{N1}) = \infty, u(e_{E2}) = \infty, u(e_{S2}) = \infty, u(e_{W2}) = \infty, u(e_{N2}) = \infty,$

(see Fig. 9.11),

- a two block vector subtracter: $TS_s = (S_s, E_s, T_s, I_s)$, where

  - $S_s = \{s_{ph1}, s_{ph2}, s_{aS1}, s_{aW1}, s_{aN1}, s_{aE2}, s_{aS2}, s_{aN2}, s_{aF12}\}$,
  - $E_s = \{e_{E1}, e_{S1}, e_{W1}, e_{N1}, e_{E2}, e_{S2}, e_{W2}, e_{N2}\}$,
  - $T_s = \{(s_{ph1}, e_{E1}, s_{aF12}),\ (s_{ph1}, e_{S1}, s_{aS1}),\ (s_{ph1}, e_{W1}, s_{aW1}),\ (s_{ph1}, e_{N1}, s_{aN1}),$
    $(s_{ph2}, e_{E2}, s_{aE2}), (s_{ph2}, e_{S2}, s_{aS2}), (s_{ph2}, e_{W2}, s_{aF12}), (s_{ph2}, e_{N2}, s_{aN2})\}$,
  - $I_s = \{s_{ph1}, s_{ph2}\}$,
  - and $l(e_{E1}) = t_{E1},\ \ l(e_{S1}) = t_{S1},\ \ l(e_{W1}) = t_{W1},\ \ l(e_{N1}) = t_{N1},\ \ l(e_{E2}) = t_{E2},$
    $l(e_{S2}) = t_{S2}, l(e_{W2}) = t_{W2}, l(e_{N2}) = t_{N2}, u(e_{E1}) = \infty, u(e_{S1}) = \infty, u(e_{W1}) =$
    $\infty, u(e_{N1}) = \infty, u(e_{E2}) = \infty, u(e_{S2}) = \infty, u(e_{W2}) = t^*_{W2}, u(e_{N2}) = \infty,$

(see Fig. 9.12).

In case of a two block vector adder, a special role is played by the state s_{aF12}. This state represents a situation when two plasmodia are fused in the attractant $attr_{E_1}/attr_{W_2}$ and the addition takes place. It is worth noting that:

- in case of a two block vector adder, the addition takes place if $t > max(t_{E1}, t_{W1})$,
- in case of a two block vector subtracter, the addition takes place if $max(t_{E1}, t_{W1}) < t \le t^*_{W2}$, and the subtraction takes place if $t > t^*_{W2}$.

In Figs. 9.13 and 9.14 Petri net models of a two block vector adder and a two block vector subtracter, respectively, are shown. Formally, we have:

- a two block vector adder: $MPN_a = \{Pl_a, Tr_a, Arc_a, w_a, m_a\}$, where:

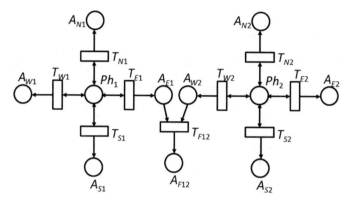

Fig. 9.13 A Petri net model of a two block vector adder

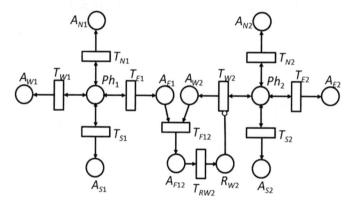

Fig. 9.14 A Petri net model of a two block vector subtracter

– $Pl_a = \{Ph_1, Ph_2, A_{E1}, A_{S1}, A_{W1}, A_{N1}, A_{E2}, A_{S2}, A_{W2}, A_{N2}, A_{F12}\}$,
– $Tr_a = \{T_{E1}, T_{S1}, T_{W1}, T_{N1}, T_{E2}, T_{S2}, T_{W2}, T_{N2}, T_{F12}\}$,
– $Arc_a = Arc_O^a \cup Arc_I^a$ such that
  · $Arc_O^a = \{(Ph_1, T_{E1}), (T_{E1}, Ph_1), (T_{E1}, A_{E1}), (Ph_1, T_{S1}), (T_{S1}, Ph_1),$
    $(T_{S1}, A_{S1}), (Ph_1, T_{W1}), (T_{W1}, Ph_1), (T_{W1}, A_{W1}), (Ph_1, T_{N1}), (T_{N1}, Ph_1),$
    $(T_{N1}, A_{N1}), (Ph_2, T_{E2}), (T_{E2}, Ph_2), (T_{E2}, A_{E2}), (Ph_2, T_{S2}), (T_{S2}, Ph_2),$
    $(T_{S2}, A_{S2}), (Ph_2, T_{W2}), (T_{W2}, Ph_2), (T_{W2}, A_{W2}), (Ph_2, T_{N2}), (T_{N2}, Ph_2),$
    $(T_{N2}, A_{N2}), (A_{E1}, T_{F12}), (A_{W2}, T_{F12}), (T_{F12}, A_{F12})\}$,
  · $Arc_I^a = \emptyset$,
– $w_a(arc) = 1$ for all $arc \in Arc_a$,
– and $m_a(p) = 0$ for each $p \in Pl_a$,

  (see Fig. 9.13),

• a two block vector subtracter: $MPN_s = \{Pl_s, Tr_s, Arc_s, w_s, m_s\}$, where:

  – $Pl_s = \{Ph_1, Ph_2, A_{E1}, A_{S1}, A_{W1}, A_{N1}, A_{E2}, A_{S2}, A_{W2}, A_{N2}, A_{F12}, R_{W2}\}$,

- $Tr_s = \{T_{E1}, T_{S1}, T_{W1}, T_{N1}, T_{E2}, T_{S2}, T_{W2}, T_{N2}, T_{F12}, T_{RW2}\}$,
- $Arc_s = Arc_O^s \cup Arc_I^s$ such that
  - $Arc_O^s = \{(Ph_1, T_{E1}),\ (T_{E1}, Ph_1),\ (T_{E1}, A_{E1}),\ (Ph_1, T_{S1}),\ (T_{S1}, Ph_1),$
    $(T_{S1}, A_{S1}),\ (Ph_1, T_{W1}),\ (T_{W1}, Ph_1),\ (T_{W1}, A_{W1}),\ (Ph_1, T_{N1}),\ (T_{N1}, Ph_1),$
    $(T_{N1}, A_{N1}),\ (Ph_2, T_{E2}),\ (T_{E2}, Ph_2),\ (T_{E2}, A_{E2}),\ (Ph_2, T_{S2}),\ (T_{S2}, Ph_2),$
    $(T_{S2}, A_{S2}),\ (Ph_2, T_{W2}),\ (T_{W2}, Ph_2),\ (T_{W2}, A_{W2}),\ (Ph_2, T_{N2}),\ (T_{N2}, Ph_2),$
    $(T_{N2}, A_{N2}),\ (A_{F12}, T_{RW2}),\ (T_{RW2}, R_{W2})\}$,
  - $Arc_I^s = \{(R_{W2}, T_{W2})\}$,
- $w_s(arc) = 1$ for all $arc \in Arc_s$,
- and $m_s(p) = 0$ for each $p \in Pl_s$,

(see Fig. 9.14).

Thus, *Physarum* machines implement *p*-adic valued arithmetics and can calculate simple arithmetic functions performing additions and subtractions in the non-well-founded universe.

Chapter 10
The Rudiments of Physarum Games

In this chapter, we are going to propose a *context-based game theory* as an example of new mathematics which could be applied in programming different biological devices, not only the Physarum Chips. The standard game theory is reduced to usual algorithmic mathematics and can be presented as multiplicative linear logic [1]. Mathematically, it is a very simple system. The context-based game theory proposed by us is constructed on transition (non-linear) systems, where agents, who can change their past decisions, move. So, we appeal to the *non-well-founded mathematics* for coding information on behavioural systems such as plasmodia in the form of non-linear games.

The standard game theory as well as standard mathematics cannot be implemented by plasmodia as a kind of processors, because of behavioural freedom and contextuality in plasmodium decisions. Logic circuits on behavioural system need the non-well-founded mathematics which can be exemplified by context-based game theory. Thus, this new game theory is to express features of plasmodium processor. However, it can be interesting as such, too, as a new scientific branch called *bio-inspired game theory*.

10.1 Transition Systems of Plasmodia and Hybrid Actions

We know that the *Physarum polycephalum* motions are a kind of *natural transition systems*, $(States, Edges)$, where $States$ is a set of states presented by attractants and $Edges \subseteq States \times States$ is a transition of plasmodium from one attractant to another.

We can define simple actions as logic gates AND, OR, and NOT. All the transitions will be built on their compositions with n inputs. So, labelled transition systems have

© Springer International Publishing AG, part of Springer Nature 2019
A. Schumann and K. Pancerz, *High-Level Models of Unconventional Computations*, Studies in Systems, Decision and Control 159,
https://doi.org/10.1007/978-3-319-91773-3_10

been used for defining the so-called *concurrent games*, a new semantics for games [2]. Traditionally, a play of the game is formalized as a sequence of moves. This way assumes the polarization of two-person games, when in each position there is only one player's turn to move. These sequential games can hold just on models of fragments of linear logic such as multiplicative [1] or multiplicative-exponential fragments [17]. In concurrent games, players can move concurrently. On the medium of *Physarum polycephalum* we can, first, define concurrent games and, second, extend the notion of concurrent games strongly and introduce the so-called *context-based games*.

While in concurrent games we deal with a finite set of actions, in context-based games we can move concurrently as well, but the set of actions is infinite. One of the simple versions of context-based games was introduced in [90], it is called reflexive games. In that version, we use cellular automata of payoffs to define reflexive games of human beings [92, 93]. The idea is that in reflexive games, on the one hand, we try to be unpredictable for other players and, on the other hand, to predict them. In these games, the set of actions is infinite. But this infiniteness is connected rather to non-well-founded properties of basic actions [3]. They cannot be regarded as atomic so that composite actions can be obtained over them inductively. In other words, it is possible to face a hybrid action which is singular, but it is not one of the basic simple actions. It is a hybrid of them. An example of this hybrid act for *Physarum polycephalum* was considered in [98].

In [109], we performed the double-slit experiment for the slime mould and showed that while in quantum mechanics we cannot approximate single photons (also, electrons, neutrons, etc.), in the experiments with *Physarum polycephalum* we cannot approximate single acts. But we define context-based games for involving hybrid acts. These games are a generalization of concurrent games, on the one hand, and reflexive games, on the other hand. In our opinion, this form of games can be regarded as *fundamentals of bio-inspired game theory*. About bio-inspired games please see [116, 120, 127, 128]. Also, this form of game is a simple example of non-well-founded mathematics that can be applied in programming the plasmodium processors or in programming processors on any other semi-predictable behavioural systems.

According to our hypothesis [98], natural computations may be constructed on an infinite set of hybrid actions, which are built up on the set of basic simple actions that is always finite. We can program the logic circuits for the unconventional computer on programmable behaviour of *Physarum polycephalum* as a kind of context-based games, where hybrid actions for transitions are obtained on logic gates AND, OR, NOT. This approach in unconventional computing proposed by us is relatively new. In this approach, on the one hand, we consider the behaviour of one-cell organisms as intelligent and we want to obtain the new game theory on the basis of biological experiments, where fundamentals are defined on the experiments with one-cell organisms. On the other hand, we design logic circuits by using non-well-founded mathematics illustrated by context-based games and these circuits can be implemented on the medium of behaviour of different organisms from parasites [111–113] to human beings [114]. The context-based game is a generalization of incomplete information game, concurrent game, repeated game, reflexive game and some other modern approaches to games.

10.2 Concurrent Games on Slime Mould

So, the *Physarum polycephalum* plasmodium can be interpreted as transition system $\mathscr{S} = (States, Edges)$, where (i) *States* is a set of states presented by attractants occupying by the plasmodium, (ii) $Edges \subseteq States \times States$ is the set of transitions presenting the plasmodium propagation from one state to another. States will be regarded as possible payoffs for *Physarum polycephalum*. By different localizations of attractants we can manage its motions differently. These localizations combined with different intensity of attractants are stimuli for *Physarum polycephalum* motions. We can interpret these stimuli as Boolean functions on payoffs. For example, we can define the following simplest logical gates:

- The *AND gate* (*Physarum conjunction*) (see Fig. 10.1), the serial connection of contacts. Plasmodium P follows conjunction $A_1 \wedge A_2$ if and only if it reaches both attractants A_1 and A_2, i.e. the plasmodium is attracted to A_1, when it is placed in its region of influence, then it is attracted by A_2 in the region of its influence. If we have deactivated at least one of the attractants A_1, A_2, the plasmodium cannot reach conjunction $A_1 \wedge A_2$, i.e. the latter is false. *Physarum polycephalum* conjunction has two inputs A_1, A_2 and one output $A_1 \wedge A_2$.
- The *OR gate* (*Physarum disjunction*) (see Fig. 10.2), the parallel connection of contacts. Plasmodium P follows disjunction $A_1 \vee A_2$ if and only if one of the attractants A_1 or A_2 or both are occupied by the plasmodium. In case of deactivation of both attractant A_1 and attractant A_2 the plasmodium cannot start from the initial

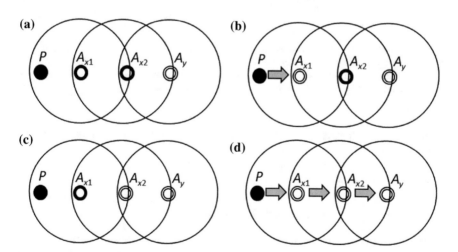

Fig. 10.1 States of the AND gate for all input combinations of A_{x1} and A_{x2} with one output A_y: **a** both attractants A_{x1} and A_{x2} are deactivated; **b** A_{x1} is activated and A_{x2} is deactivated; **c** A_{x1} is deactivated and A_{x2} is activated; **d** both A_{x1} and A_{x2} are activated. The *Physarum polycephalum* plasmodium denoted by P begins to move from the left hand-side

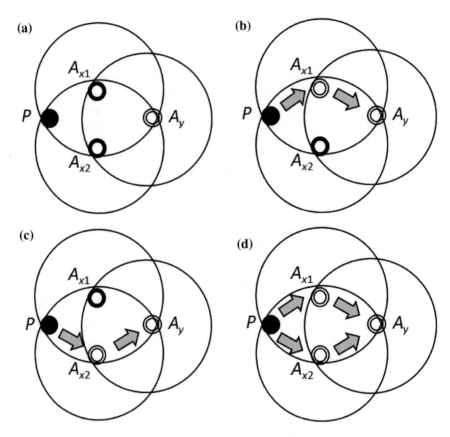

Fig. 10.2 States of the OR gate for all input combinations of A_{x1} and A_{x2} with one output A_y: **a** both attractants A_{x1} and A_{x2} are deactivated; **b** A_{x1} is activated and A_{x2} is deactivated; **c** A_{x1} is deactivated and A_{x2} is activated; **d** both A_{x1} and A_{x2} are activated. The *Physarum polycephalum* plasmodium denoted by P begins to move from the left hand-side

position. This means that $A_1 \vee A_2$ is false. *Physarum polycephalum* disjunction has two inputs A_1, A_2 and one output $A_1 \vee A_2$.

- The *NOT gate* (*Physarum negation*) (see Fig. 10.3). Plasmodium P follows negation $\neg A_1$ if and only if its behaviour is simulated by the repellent R before the attractant A_1, which avoids the plasmodium to be attracted by A_1. *Physarum polycephalum* negation has one input A_1 and one output $\neg A_1$.

Plasmodia perform logical gates to realize Boolean functions on payoffs. Games constructed on reasoning on payoffs are called *reflexive* [90]. Some of them can be simulated by means of *Physarum polycephalum* behaviours.

Hence, the propagation of *Physarum polycephalum* plasmodium is understood by us as a transition system $TS(\mathscr{PM}) = (S, E, T, S_{init})$. States S will be regarded as possible payoffs for *Physarum polycephalum*. Events E will be examined as allowed

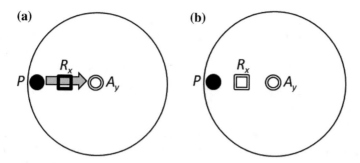

Fig. 10.3 States of the NOT gate for all input combinations with one output A_y: **a** repellent R_x is deactivated; **b** R_x is activated

moves in slime mould games. Transitions T is represented as a set of states resulting from the moves and initial states S_{init} as different players.

For possible actions in plasmodium propagations, attracting, repelling, splitting, and fusing (see Chap. 3), we can define logical operations on S as follows:

- *negation* $\neg s_x$ is true if and only if there is no s_y such that $s_y \xrightarrow{e_{yx}} s_x$ and it is false otherwise;
- *conjunction* $s_x \wedge s_y = \min(s_x, s_y)$ is true if and only if for both s_x and s_y there is s_z such that $s_z \xrightarrow{e_{zx}} s_x$ **and** $s_z \xrightarrow{e_{zy}} s_y$ and it is false otherwise;
- *disjunction* $s_x \vee s_y = \max(s_x, s_y)$ is true if and only if for both s_x and s_y there is s_z such that $s_z \xrightarrow{e_{zx}} s_x$ **or** $s_z \xrightarrow{e_{zy}} s_y$ and it is false otherwise;

If reflexive games are constructed by a finite set of actions defined as inductive compositions of logic gates presented above, then these games are called *concurrent*, where plasmodia are players:

Definition 10.1 A (finite) *concurrent game on Physarum polycephalum* is a tuple $\mathcal{G} = (States, Agt, Act_n, Mov^n, Tab^n, (\preceq_A)_{A \in Agt})$, where

- *States* is a (finite) set of *states* presented by attractants occupying by the plasmodium;
- $Agt = \{1, \ldots, k\}$ is a finite set of *players* presented by different active zones of plasmodium;
- Act_n is a non-empty set of *actions* presented by logic gates or their inductive combinations with n inputs and one output, an element of Act_n^{Agt} is called a *move*;
- $Mov^n : States^n \times Act_n^{Agt} \to 2^{Act} \setminus \{\emptyset\}$ is a mapping indicating the *available* sets of actions to a given player in a given set of states, $n > 0$ is said to be a radius of plasmodium actions, a move $m_{Agt}^n = (m_A^n)_{A \in Agt}$ is legal at $(s_1, \ldots, s_n)$ if $m_A^n \in Mov^n(\mathbf{s}, A)$ for all $A \in Agt$, where $\mathbf{s} = (s_1, \ldots, s_n)$;
- $Tab^n : States^n \times Act_n^{Agt} \to States$ is the transition table which associates, with a given set of states and a given move of the players, the set of states resulting from that move;

- for each $A \in Agt$, $\preceq_A$ is a preorder (reflexive and transitive relation) over $States^\omega$, called the *preference relation* of player A, indicating the intensity of attractants; for each $\pi, \pi' \in States^\omega$, by $\pi \preceq_A \pi'$ we denote that π' is *at least as good as* π for A and when it is not $\pi \preceq_A \pi'$, we say that A *prefers* π over π'.

Assume there is a game G and its states are formulated as some propositions, whose set is denoted by $Prop$.

Definition 10.2 A concurrent game on *Physarum polycephalum* $\mathscr{G}$ is a *model* $\mathscr{M}$ for the game G if $\mathscr{M} = (\mathscr{G}, \varphi)$, where $\varphi : States \to \mathscr{P}(Prop)$ is a labelling function such that it labels the states in $\mathscr{G}$ by proposition symbols from the set $Prop$ of the game G.

Usually, for each player (plasmodium) named $1, \ldots, k$, where $k = |Agt|$, we have a separate space of attractants. However, the space can be joint, too. The point is that if two players (plasmodia) move to the same state, then their transitions from this state are the same. In order to avoid this property, we should design different moves of different players in different spaces.

The set of *outcomes* $Out_\mathscr{G}(s)$ of the concurrent game $\mathscr{G}$ from the state s is a set of all infinite paths $s_0 s_1 \ldots \in States^\omega$ such that $s_0 = s$ and for all $j > 0$, there exists a move $m \in \prod_{k=1}^{|Agt|} Mov(s_j, k)$ and $Tab(s_j, m) = s_{j+1}$. The *set of all outcomes* is as follows: $Out_\mathscr{G} = \bigcup_{s' \in States} Out_\mathscr{G}(s)$. The set of *histories* $Hist_\mathscr{G}(s)$ starting in s is a set of all finite paths $s_0 s_1 \ldots s_l$ such that $s_0 = s$ and there exists $\sigma \in Out_\mathscr{G}(s)$ which starts with $s_0 s_1 \ldots s_l$. The set $Hist_\mathscr{G} = \bigcup_{s' \in States} Hist_\mathscr{G}(s)$ is said to be a *set of all histories*.

A *strategy* of a player j in $\mathscr{G}$ is a mapping $strat_j : Hist_\mathscr{G} \to Act$ such that for any history $\sigma \in Hist_\mathscr{G}$ it is true that $strat_j(\sigma) \in Mov(last(\sigma), j)$, where $last(\sigma)$ denotes the last state in the finite path σ. In other words, a strategy $strat_j(\sigma)$ is the choice of legal action in the last state of the history σ which was observed by player j. If we have one space for all players and they have the same last state $last(\sigma)$, then for all of them the next legal action will be the same, too. So, in this case histories of players are unimportant for choosing the next action, i.e. since that $last(\sigma)$ their histories will be the same.

Andrew Adamatzky and Martin Grube have performed some experiments showing that there are cases when sets of strategies for players in the same space are always disjoint. Let us suppose that we have only two agents. The first is presented by a usual *Physarum polycephalum* plasmodium. The second by its modification called a *Badhamia utricularis* plasmodium. *Physarum polycephalum* grows definitely faster than *Badhamia utricularis* and overtakes more flakes at the same time than the latter (see Fig. 10.4). Only if the inoculum was "fatter" for *Badhamia utricularis*, it might grow faster. Moreover, if the invasive growth front of *Badhamia utricularis* is well nourished by oat it easily overgrows the opposing tube system of *Physarum polycephalum*. So, at the microscopic level we can find out that in most observations *Physarum polycephalum* could grow into branches of *Badhamia utricularis*, while *Badhamia utricularis* could grow over *Physarum polycephalum* strands. We can see that somehow *Physarum polycephalum* feeds on small branches of *Badhamia*

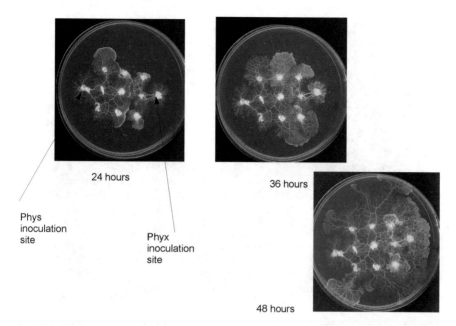

24 hours

36 hours

Phys
inoculation
site

Phyx
inoculation
site

48 hours

Fig. 10.4 The experiment (performed by Andrew Adamatsky) with two agents: fronts of growing pseudopodia of *Physarum polycephalum* (Phys for short) and *Badhamia utricularis* (PhyX for short). The picture was taken from [127]

utricularis (see Fig. 10.5). Thus, in case of *Physarum polycephalum* and *Badhamia utricularis* we observe a competition in the small branches. For them the sets of strategies are disjoint. They never meet the same states.

On the basis of competitions between *Physarum polycephalum* and *Badhamia utricularis*, we can study simplest biological forms of *zero-sum games*.

A *strategy for several players* A is defined as the following tuple $(strat_j)_{j \in A}$ of strategies for all players of A. The set of all outcomes if the players in A follow the strategy $strat_A = (strat_j)_{j \in A}$ is denoted by $Out_{\mathcal{G}}(s, strat_A)$. All possible outcomes if the players in A obey $strat_A$ is denoted by $Out_{\mathcal{G}}(strat_A) = \bigcup_{s \in States} Out_{\mathcal{G}}(s, strat_A)$. A strategy for Agt: $(strat_j)_{j \in Agt}$ is called a *strategy profile*.

We will say that a strategy $strat_A$ is *memoryless* for players of A at a state s if they choose their joint action based only on s as the last state of play. This holds if they meet s in a joint space, simultaneously. Notice that in a game with two players presented by *Physarum polycephalum* plasmodium 1 and *Badhamia utricularis* plasmodium 2 in a joint space, their strategies $strat_{Agt=\{1,2\}}$ cannot be memoryless at any time.

Let us take a move m_{Agt} and an action m' for some player B. Assume, the move n_{Agt} with $n_A = m_A$ when $A \neq B$ and $n_B = m'$ is denoted by $m_{Agt}[B \to m']$. Then a Nash equilibrium for concurrent games is defined as follows (for more details on equilibria in concurrent games, see [20–22]):

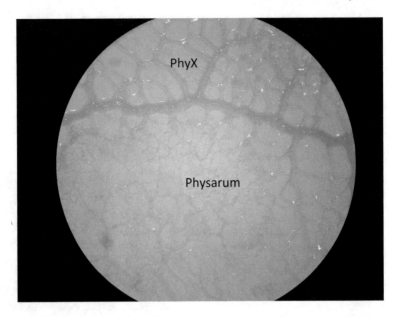

Fig. 10.5 The experiment (performed by Andrew Adamatsky) with two agents: *Physarum poly-cephalum* (Physarum for short) could grow into branches of *Badhamia utricularis* (PhyX for short). The picture was taken from [127]

Definition 10.3 Let $\mathcal{G}$ be a concurrent game with preference relation $(\preceq_A)_{A \in Agt}$ and let s be a state of $\mathcal{G}$. A *Nash equilibrium* of $\mathcal{G}$ from s is a strategy profile $strat_{Agt}$ such that $Out(s, strat_{Agt}[B \rightarrow strat']) \preceq_B Out(s, strat_{Agt})$ for all players $B \in Agt$ and all strategies $strat'$ of B.

Notice that concurrent games can be implemented in the plasmodium behaviour only with low accuracy, because concurrent games do not assume possibilities to change past decisions, i.e. the set of possible actions is always finite. Nevertheless, the set of actions for the plasmodium is infinite.

10.3   Context-Based Games on Slime Mould

Let us consider the following thought experiment as counterexample showing that the set of actions for the plasmodium is infinite in principle. This means that concurrent games can be only approximated on the medium of slime mould, i.e. plasmodia follow concurrent games only uncertainly and probabilistically. Assume that the labelled transition system for the plasmodium consists just of one action presented by one neighbour attractant. The plasmodium is expected to propagate a protoplasmic tube towards this attractant. Now, let us place a barrier with one slit in front of the plasmodium. Because of this slit, the plasmodium can be propagated according to the shortest distance between two points and in this case the plasmodium does not pay attention on the barrier. However, sometimes the plasmodium can evaluate the same barrier as a repellent for any case and it gets round the barrier to reach the attractant according to the longest distance. So, even if the environment conditions change a little bit, the behaviour changes, too. The plasmodium is very sensible to the environment.

Thus, simple actions of *Physarum* plasmodia cannot be regarded as atomic so that composite actions can be obtained over them inductively. In the labelled transition system with only one stimulus presented by one attractant, a passable barrier can be evaluated as a repellent 'for any case'. Therefore the transition system with only one stimulus and one passable barrier may have the following three simple actions: (i) pass trough, (ii) avoid from the left, (iii) avoid from the right. But in essence, we deal only with one stimulus and, therefore, with one action, although this action has the three modifications defined above. Simple actions which have modifications depending on the environment are called *hybrid*. The problem is that the set of actions in any labelled transition systems must consist of the so-called atomic actions—simple actions that have no modifications.

Let us show in examples, how we can extend the definition of concurrent games up to the case of context-based games, where actions can be hybrid.

Example 10.1 Let $\mathscr{G} = (States, Agt, Act_n, Mov^n, Tab^n, (\preceq_A)_{A \in Agt})$ and $n = 9$ (i.e. our game is 10-adic valued), $Agt = \{A_1, A_2\}$, $States = \{s_1, s_2, \ldots, s_m\}$ ($m \geq p = 10$), $Act_n = \{\max, \min\}$. Assume that A_1 follows only max at all legal moves and A_2 follows only min at his legal moves. This means that in transition system $TS(\mathscr{PM})$, $E_{A_i} = \{(\mathbf{s}, \mathbf{s}') \in S^9 \times S^9 : Tab^9(\mathbf{s}, m_{A_i}^9) = \mathbf{s}'\}$, where $i = 1, 2$, $m_{A_1}^9 = \max$, and $m_{A_2}^9 = \min$. This game can be illustrated in the cellular-automatic form with the neighbourhood $|N| = 8$, see [90]. Let us take cells belonging to the set $\mathbf{Z}^2$, therewith each cell takes its value in *States*. Let transitions depend on local transition rule $\delta: States^9 \to States$ that transforms states of cells taking into account states of 8 neighbour cells. Each step of dynamics is fixed by discrete time $t = 0, 1, 2, \ldots$ At the moment t, the configuration of the whole system (or the global state) is given by the mapping x^t from $\mathbf{Z}^2$ into *States*, and the evolution is the sequence $x^0 x^1 x^2 x^3 \ldots$ defined as follows: $x^{t+1}(z) = \delta(x^t(z), x^t(z + \alpha_1), x^t(z + \alpha_2), \ldots, x^t(z + \alpha_8))$, where $(\alpha_1, \alpha_2, \ldots, \alpha_8)$ are neighbours of z.

Assume that at each move an occupied attractant from *States* is denoted by 1 and an unoccupied attractant is denoted by 0. Let us suppose, at $t = 0$ we have the following states, where, given (s_i, s_j), s_i means a state for A_1 and s_j means a state for A_2:

(0, 1)	(0, 0)	(1, 1)	(0, 1)	(1, 0)
(1, 1)	(1, 1)	(1, 0)	(1, 0)	(1, 1)
(0, 1)	(1, 0)	(1, 1)	(1, 0)	(0, 0)
(0, 0)	(1, 0)	(0, 0)	(1, 1)	(1, 0)
(0, 1)	(1, 0)	(1, 1)	(1, 1)	(0, 0)

Then at $t = 1$ we obtain the following states:

$(1,0)$	$(1,0)$	$(1,0)$	$(1,0)$	$(1,0)$
$(1,0)$	$(1,0)$	$(1,0)$	$(1,0)$	$(1,0)$
$(1,0)$	$(1,0)$	$(1,0)$	$(1,0)$	$(1,0)$
$(1,0)$	$(1,0)$	$(1,0)$	$(1,0)$	$(1,0)$
$(1,0)$	$(1,0)$	$(1,0)$	$(1,0)$	$(1,0)$

Hence, we deal with the set of actions presented by logical operations over $States$ or their inductive combinations which give n outputs for n inputs and we take into account time t.

Example 10.2 Our other example is as follows. Let

$$\mathscr{G} = (States_t, Agt, Act_{t,n}, Mov_t^n, Tab_t^n, (\preceq_A)_{A \in Agt}),$$

where

- $States_t$ means states for different time $t = 0, 1, \ldots$;
- Agt is a finite set of players;
- $Act_{t,n}$ is an infinite set of actions presented by logical operations or their inductive combinations with n inputs and n outputs, these actions can be applicable only to states of time t and give states of time $t + 1$;
- Mov_t^n is a set of legal moves at time $t = 0, 1, \ldots$;
- $Tab_t^n \colon States_t^n \times Act_{t,n}^{Agt} \to States_{t+1}^n$ is the transition table which associates, with a given set of states and a given move of the players, the set of states resulting from that move;
- for each $A \in Agt$, $\preceq_A$ is a preorder over $States^\omega$, called the *preference relation* of player A.

Suppose that $n = 9$ (i.e. the game is 10-adic valued still), $Agt = \{A_1, A_2\}$, $States = \{s_1, s_2, \ldots, s_m\}$ ($m \geq p = 10$) such that each state has its value in $\{0, 1\}$ (i.e. it is occupied or not), $Act_{t,9} = \{\max \Rightarrow \min, \min \Rightarrow \max\}$. Assume that A_1 follows the rule: $\max\{$the states of A_1 at $t\} \Rightarrow \min\{$the states of A_2 at $t\}$, and A_2 follows the rule: $\min\{$the states of A_2 at $t\} \Rightarrow \max\{$the states of A_1 at $t\}$. This means that in an appropriate transition system $TS(\mathscr{PM})$, $E_{A_i} = \{(\mathbf{s}, \mathbf{s}') \in S_t^9 \times S_{t+1}^9 \colon Tab^9(\mathbf{s}, m_{A_i}^9) = \mathbf{s}'\}$, where $i = 1, 2$, $m_{A_1}^9 = \max\{$the states of A_1 at $t\} \Rightarrow \min\{$the states of A_2 at $t\}$, and $m_{A_2}^9 = \min\{$the states of A_2 at $t\} \Rightarrow \max\{$the states of A_1 at $t\}$.

In the cellular-automatic form, at $t = 0$:

(0, 1)	(0, 0)	(1, 1)	(0, 1)	(1, 0)
(1, 1)	(1, 1)	(1, 0)	(1, 0)	(1, 1)
(0, 1)	(1, 0)	(1, 1)	(1, 0)	(0, 0)
(0, 0)	(1, 0)	(0, 0)	(1, 1)	(1, 0)
(0, 1)	(1, 0)	(1, 1)	(1, 1)	(0, 0)

At $t = 1$:

(1, 0)	(0, 1)	(0, 1)	(0, 1)	(0, 1)
(0, 1)	(0, 1)	(0, 1)	(0, 1)	(0, 1)
(0, 1)	(0, 1)	(0, 1)	(0, 1)	(0, 1)
(0, 1)	(0, 1)	(0, 1)	(0, 1)	(0, 1)
(0, 1)	(0, 1)	(0, 1)	(0, 1)	(0, 1)

We have just shown that we can involve actions which are different for all $t = 0, 1, \ldots$ and are mutually dependent. These actions are hybrid.

Example 10.3 Let us consider an example, when $n = 1$ (i.e. the game is 2-adic valued) and $Agt = \{A_1, A_2\}$ in $\mathscr{G}$, to demonstrate that the set of actions for slime mould games can be uncountably infinite.

Assume that agent A_1 moves from the state s_1^t to the state s_1^{t+1}, $t = 0, 1, \ldots$ and agent A_2 moves from the state s_2^t to the state s_2^{t+1}, $t = 0, 1, \ldots$ by the following transition rules: $s_1^{t+1} = (s_1^t \Rightarrow s_2^t)$ and $s_2^{t+1} = (s_2^t \Rightarrow s_1^t)$. Then we obtain the following infinite streams: $(s_1^0 \Rightarrow s_2^0, s_1^1 \Rightarrow s_2^1, s_1^2 \Rightarrow s_2^2, \ldots)$ and $(s_2^0 \Rightarrow s_1^0, s_2^1 \Rightarrow s_1^1, s_2^2 \Rightarrow s_1^2, \ldots)$. Let us pay attention that the stream $(s_1^0 \Rightarrow s_2^0, s_1^1 \Rightarrow s_2^1, s_1^2 \Rightarrow s_2^2, \ldots)$ (resp. $(s_2^0 \Rightarrow s_1^0, s_2^1 \Rightarrow s_1^1, s_2^2 \Rightarrow s_1^2, \ldots)$) may be understood as an infinite propositional formula $(((s_1^0 \Rightarrow s_2^0) \Rightarrow s_2^1) \Rightarrow s_2^2) \Rightarrow \ldots$ (resp. $(((s_2^0 \Rightarrow s_1^0) \Rightarrow s_1^1) \Rightarrow s_1^2) \Rightarrow \ldots$). Both formulas are mutually dependent and they cannot be presented as linear sequence (inductive composition). Hence, we have a 2-adic valued formula presented by two infinite mutually dependent propositional formulas. This formula is called *non-well-founded* [3] or *hybrid* [98]. Also, we can show that any non-well-founded formula of radius $n = |Agt|$ can be formulated as n infinite mutually dependent propositional formulas. In concurrent games, there is no reflexion of players. They do not pay attention, which actions are involved in transitions by others. A non-well-founded action of radius $n = |Agt|$ means that each player of Agt coordinates his action with others by using their reasoning in the form he can foresee, maybe wrongly. Reflexive games are considered in [90, 92, 93]. Preference relations for reflexive games are examined in [114]. They are strong extensions of concurrent games.

In the case of slime mould, non-well-founded actions are not results of predictions of others as in the case of human reflexive games [93]. The matter is that the one plasmodium can follow some non-well-founded actions [98], since for $n \geq 2$ inputs there is an uncertainty, which logic gates with n inputs are involved in fact. Therefore we can consider all possible logic gates with n inputs. These possible gates cannot exceed the number n. So, maximally, we can have only n outputs.

Definition 10.4 A (finite) *context-based game on Physarum polycephalum* is a tuple $\mathcal{G} = (States_t, Agt, Act_{t,n}, Mov_t^n, Tab_t^n, (\preceq_A)_{A \in Agt})$, where

- $States_t$ is a (finite) set of *states* presented by attractants occupying by the plasmodium at time $t = 0, 1, 2, \ldots$;
- $Agt = \{1, \ldots, k\}$ is a finite set of *players* presented by different active zones of plasmodium;
- $Act_{t,n}$ is a non-empty set of *hybrid actions* with radius n at $t = 0, 1, 2, \ldots$, an element of $Act_{t,n}^{Agt}$ is called a *move* at time $t = 0, 1, 2, \ldots$;
- $Mov_t^n : States_t^n \times Act_{t,n}^{Agt} \to 2^{Act} \setminus \{\emptyset\}$ is a mapping indicating the *available* sets of actions to a given player in a given set of states, $n > 0$ is said to be a radius of plasmodium actions, a move $m_{Agt}^n = (m_A^n)_{A \in Agt}$ is legal at $(s_1, \ldots, s_n)$ if $m_A^n \in Mov^n(\mathbf{s}, A)$ for all $A \in Agt$, where $\mathbf{s} = (s_1, \ldots, s_n)$;
- $Tab_t^n : States_t^n \times Act_{t,n}^{Agt} \to States_{t+1}^n$ is the transition table which associates, with a given set of states at t and a given move of the players at t, the set of states at $t + 1$ resulting from that move;
- for each $A \in Agt$, $\preceq_A$ is a preorder (reflexive and transitive relation) over $States^\omega$, called the *preference relation* of player A, indicating the intensity of attractants.

It is worth noting that context-based games are *massively parallel* [85–88, 115].

All other notions such as outcomes, histories, and strategies are defined in the way of the previous section. We can prove a statement that for any concurrent game $\mathcal{G}$ on the medium of slime mould, there is an appropriate context-based game as a greatest fixed point for all uncertain modifications of $\mathcal{G}$ in experiments with plasmodia. This statement allows us to build logic circuits on the slime mould using context-based game notions.

In the concurrent games as well as in the context-based games we can use rough sets for defining strategies. Let $\Omega^t = \{\omega_1^t, \omega_2^t, \ldots, \omega_k^t\}$ be a set of all *nearest strategies* at t for all agents $Agt = \{1, 2, \ldots, k\}$, i.e. strategies performed only one time at the actual time step t. Suppose, $\mathcal{N} = \{N_{\omega_1^t}, N_{\omega_2^t}, \ldots, N_{\omega_k^t}\}$ denotes a family of payoffs corresponding to nearest strategies such that $N_{\omega_i^t} \subset States$ represents the states obtained by player $i = \overline{1, k}$ by applying strategy ω_i^t at the actual time step t.

If for each i, ω_i^t gives only a singleton $N_{\omega_i^t}$, i.e. $card(N_{\omega_i^t}) = 1$, the game is called *concurrent*. If for some i, ω_i^t gives $N_{\omega_i^t}$ such that $card(N_{\omega_i^t}) > 1$, the game is called *context-based*.

For each state $s \in N_{\omega_1^t} \cup N_{\omega_2^t} \cup \cdots \cup N_{\omega_k^t}$, we define its p-adic valued inter-region neighbourhood at t:

$$IRN_t^p(s) = \{s' : (s, s') \in E \land \underset{\omega \in \Omega^t}{\exists} (s' \in N_\omega \land s \notin N_\omega)\}.$$

The cardinal number $card(IRN_t^p(s)) \leq p$.

We assume that each player can change his/her strategies at each new step t. It means that we deal with a transition $\omega_i^t \to \omega_i^{t+1}$.

The p-adic valued lower approximation $\underline{IRN}_{t+1}^P(\omega_i^t \to \omega_i^{t+1})$ of the strategy change at $t+1$ is defined as follows:

$$\underline{IRN}_{t+1}^P(\omega_i^t \to \omega_i^{t+1}) = \{s \in N_{\omega_i^t} : IRN_t^P(s) \neq \emptyset \wedge IRN_t^P(s) \subseteq N_{\omega_i^{t+1}}\}.$$

The p-adic valued upper approximation $\overline{IRN}_{t+1}^P(\omega_i^t \to \omega_i^{t+1})$ of the strategy change at $t+1$ is defined thus:

$$\overline{IRN}_{t+1}^P(\omega_i^t \to \omega_i^{t+1}) = \{s \in N_{\omega_i^t} : IRN^P(s) \cap N_{\omega_i^{t+1}} \neq \emptyset\}.$$

The *intentionality of player i* by a strategy change at $t+1$ is defined in the following manner:

$$\alpha_{IRN}(\omega_i^t \to \omega_i^{t+1}) = \frac{card(\underline{IRN}_{t+1}^P(\omega_i^t \to \omega_i^{t+1}))}{card(\overline{IRN}_{t+1}^P(\omega_i^t \to \omega_i^{t+1}))}.$$

We see that

$$0 \leq \alpha_{IRN}(\omega_i^t \to \omega_i^{t+1}) \leq 1.$$

The situation $\alpha_{IRN}(\omega_i^t \to \omega_i^{t+1}) = 0$ means that $card(\underline{IRN}_{t+1}^P(\omega_i^t \to \omega_i^{t+1})) = 0$, i.e. it means that new payoffs at $t+1$, if they take place, are obtained not intentionally. The case $\alpha_{IRN}(\omega_i^t \to \omega_i^{t+1}) = 1$ means that $card(\underline{IRN}_{t+1}^P(\omega_i^t \to \omega_i^{t+1})) = card(\overline{IRN}_{t+1}^P(\omega_i^t \to \omega_i^{t+1}))$, i.e. it means that all the new payoffs at $t+1$ are obtained intentionally. Hence, the measure $\alpha_{IRN}(\omega_i^t \to \omega_i^{t+1})$ tells us about intentionality of player i in moving from t to $t+1$.

Now, let us define the *intentionality of player i* through the whole game, assuming that it is infinite. Let

$$\alpha_{IRN}(i) = \frac{\sum_{t=0}^{\infty} card(\underline{IRN}_{t+1}^P(\omega_i^t \to \omega_i^{t+1})) \cdot p^t}{\sum_{t=0}^{\infty} card(\overline{IRN}_{t+1}^P(\omega_i^t \to \omega_i^{t+1})) \cdot p^t}.$$

Then also:

$$0 \leq \alpha_{IRN}(i) \leq 1.$$

But now this measure runs over the set of p-adic integers $\mathbf{Z}_p$.

We suppose that *somebody wins* if (s)he has occupied more payoffs (attractants) at the majority steps $t \to \infty$ than each other player separately. Somebody *loses* if (s)he has occupied less payoffs (attractants) at the majority steps $t \to \infty$ than each other player separately. Let us define the same formally.

Let $\mathcal{N} = \{N_{\omega_1^t}, N_{\omega_2^t}, \dots, N_{\omega_k^t}\}$ be a family of payoffs corresponding to near-est strategies ω_i^t at the actual time step t for each player $i = \overline{1, k}$. Let $\mathcal{N}^\emptyset = \{N_{\omega_1^t}^\emptyset, N_{\omega_2^t}^\emptyset, \dots, N_{\omega_k^t}^\emptyset\}$ be a family of vacant attractants at t such that $N_{\omega_i^t}^\emptyset = \{s_{t+1} : \underset{s_t \in N_{\omega_i^t}}{\forall} (s_t, s_{t+1}) \in E \wedge \underset{\omega \in \Omega^t}{\forall} (s_{t+1} \notin N_\omega)\}$, i.e. it is a set of all accessible attractants for player i at time t which can be occupied at $t + 1$ from the set $N_{\omega_i^t}$.

The *probability of winning* for i is defined as follows:

$$Win(i) = \frac{\sum_{t=0}^{\infty} card(N_{\omega_i^{t+1}}) \cdot p^t}{\sum_{t=0}^{\infty} card(N_{\omega_i^t}^\emptyset) \cdot p^t}.$$

Player i wins if $Win(i) \geq Win(j)$ for any player $j \in (Agt - \{i\})$. Player i loses if $Win(i) \leq Win(j)$ for any player $j \in (Agt - \{i\})$.

In case $N_{\omega_1^t}, N_{\omega_2^t}, \dots, N_{\omega_k^t}$ are pairly disjoint, the game is being carried out inde-pendently of competitor's strategies—each player plays in a parallel manner without intercommunication. Suppose now that $N_{\omega_1^t} \cap N_{\omega_2^t} \cap \cdots \cap N_{\omega_k^t} = N_t^c \neq \emptyset$.

The lower approximation $\underline{N_{\omega_i^{t+1}}}$ of payoffs for player i at $t + 1$:

$$\underline{N_{\omega_i^{t+1}}} = \{s \in N_{\omega_i^t} : N_{\omega_i^{t+1}} \neq \emptyset \wedge N_{\omega_i^{t+1}} \subseteq N_{t+1}^c\}.$$

The upper approximation $\overline{N_{\omega_i^{t+1}}}$ of payoffs for player i at $t + 1$:

$$\overline{N_{\omega_i^{t+1}}} = \{s \in N_{\omega_i^t} : N_{\omega_i^{t+1}} \cap N_{t+1}^c \neq \emptyset\}.$$

The *probability of winning in competitions for attractants* for i is defined thus:

$$Win_c(i) = \frac{\sum_{t=0}^{\infty} card(\underline{N_{\omega_i^{t+1}}}) \cdot p^t}{\sum_{t=0}^{\infty} card(\overline{N_{\omega_i^{t+1}}}) \cdot p^t}.$$

Player i wins in competitions for food if $Win_c(i) \geq Win_c(j)$ for any player $j \in (Agt - \{i\})$. Player i loses if $Win_c(i) \leq Win_c(j)$ for any player $j \in (Agt - \{i\})$.

Hence, in context-based games we have the following main features:

- the game can be infinite and its measures are set up by rough sets with values running over p-adic integers $\mathbf{Z}_p$;
- the game is concurrent if each move for each player gives only one payoff, other-wise the game is context-based;
- each player can change his/her strategy at each time step t;
- if the set of payoffs for agents i and j are intersected at t, it means that the strategies of i and j are intersected also at t.

10.4 *p*-Adic Valued Probabilities and Fuzziness

Now we can supplement the *Physarum* language (see Chap. 7) with instructions enabling us to determine (at the simulation stage) possible properties of experiments in terms of the probability space:

- *setTimeStart*—setting a time start from which the experiment starts, $t = 0, 1, 2, \ldots, n,$
- *setTimeEnd*—setting a time end when the experiment stops, $t = 0, 1, 2, \ldots, \infty,$
- *getNeighCard*—getting a cardinality number of neighbouring attractants for a given attractant at the given time start and time end,
- *getAccessCard*—getting a cardinality number of attractants accessible for a given attractant by protoplasmic tubes at the given time start and time end.

Instructions for the simulation stage are preceded with $. Let us consider a simple timed transition system given earlier.

If we add the following instructions to the code:

```
$setTimeStep(0);
$setTimeEnd(10);
$getNeighCard(s2);
$getAccessCard(s2);
```

we obtain the following 3-adic streams:

- 2 2 2 2 2 2 2 2 2 2 2 for getting a cardinality number of neighbouring attractants for A_{s_2},
- 2 2 2 2 2 2 1 1 1 1 1 for getting a cardinality number of attractants accessible for A_{s_2}.

Thus, if we have $p - 1$ neighbour attractants for A_{s_2}, we deal with *p*-adic streams. If *setTimeStep*(0) and *setTimeEnd*(∞) we deal with infinite *p*-adic streams. All these streams including both finite and infinite can be identified with *p*-adic integers

$$n = \alpha_0 + \alpha_1 \cdot p + \cdots + \alpha_t \cdot p^t + \cdots = \sum_{t=0}^{\infty} \alpha_t \cdot p^t,$$

where $\alpha_t \in \{0, 1, \ldots, p - 1\}$, $\forall t \in N$. This number sometimes has the following notation: $n = \ldots \alpha_3 \alpha_2 \alpha_1 \alpha_0$, where α_t can be interpreted as a value of α at time step $t = 0, 1, 2, \ldots, \infty$. We have used the latter notation in our example. The set of *p*-adic integers is denoted by $\mathbf{Z}_p$.

The set $\mathbf{Z}_p$ cannot be linearly ordered, but there are many possibilities to define a partial ordering relation. For example, we can assume that (i) for any finite *p*-adic integers $\sigma, \tau \in N$, we have $\sigma \leq \tau$ in N iff $\sigma \leq \tau$ in $\mathbf{Z}_p$; (ii) each finite *p*-adic integer $n = \ldots \alpha_3 \alpha_2 \alpha_1 \alpha_0$ (i.e. such that $\alpha_i = 0$ for any $i > j$) is less than any infinite number τ, i.e. $\sigma < \tau$ for any $\sigma \in N$ and $\tau \in \mathbf{Z}_p \setminus N$; (iii) each infinite *p*-adic integer σ is less, than p-adic integer τ iff $\sigma_t \leq \tau_t$ for all $t = 0, 1, 2, \ldots$ Let us denote this ordering relation by $\mathcal{O}_{\mathbf{Z}_p}$. We can see that there exist p-adic integers, which are incompatible by $\mathcal{O}_{\mathbf{Z}_p}$.

Now we can define sup and inf digit by digit. Then if $\sigma \leq \tau$, so $\inf(\sigma, \tau) = \sigma$ and $\sup(\sigma, \tau) = \tau$. The greatest p-adic integer according to our definition is $-1 = \ldots xxxxxx$, where $x = p - 1$, and the smallest is $0 = \ldots 00000$.

Let us define the Boolean operations on attractants $A_{s_i}, A_{s_j}, \ldots$ so that

$$getNeighCard(si \cap sj):: =$$

$$\inf(getNeighCard(si), getNeighCard(sj));$$

$$getAccessCard(si \cap sj):: =$$

$$\inf(getAccessCard(si), getAccessCard(sj));$$

$$getNeighCard(si \cup sj):: =$$

$$\sup(getNeighCard(si), getNeighCard(sj));$$

$$getAccessCard(si \cup sj)$$

$$:: = \inf(getAccessCard(si), getAccessCard(sj));$$

$$getNeighCard(\neg si):: = -1 - getNeighCard(si);$$

$$getAccessCard(\neg si):: = -1 - getAccessCard(si).$$

Let Ω^* denote all attractants both activated and deactivated at each $t = 0, 1, 2, \ldots, \infty$. It is a union of all attractants $A_{s_i}, A_{s_j}, \ldots$ at each time step. Its subsets will be denoted by $A^*, B^* \subseteq \Omega^*$.

Let us define p-adic fuzziness as follows: a p-adic fuzzy measure is a set function $F_{Z_p}(\cdot)$ defined for sets $A^*, B^* \subseteq \Omega^*$, it runs over the set $\mathbf{Z}_p$ and satisfies the following properties:

- $F_{Z_p}(\Omega^*) = -1$ and $F_{Z_p}(\emptyset^*) = 0$.
- If $A^* \subseteq \Omega^*$ and $B^* \subseteq \Omega^*$ are disjoint, i.e. $\inf(F_{Z_p}(A^*), F_{Z_p}(B^*)) = 0$, then $F_{Z_p}(A^* \cup B^*) = F_{Z_p}(A^*) + F_{Z_p}(B^*)$. Otherwise, $F_{Z_p}(A^* \cup B^*) = F_{Z_p}(A^*) + F_{Z_p}(B^*) - \inf(F_{Z_p}(A^*), F_{Z_p}(B^*)) = \sup(F_{Z_p}(A^*), F_{Z_p}(B^*))$.
- If $A^*, B^* \subseteq \Omega^*$, then $F_{Z_p}(A^* \cap B^*) = \inf(F_{Z_p}(A^*), F_{Z_p}(B^*))$.
- $F_{Z_p}(\neg A^*) = -1 - F_{Z_p}(A^*)$ for all $A^* \subseteq \Omega^*$, where $\neg A^* = \Omega^* \setminus A^*$.

A p-adic probability measure is a set function $P_{Z_p}(\cdot)$ defined for sets $A^*, B^* \subseteq \Omega^*$ thus:

- $P_{Z_p}(A^*) = -F_{Z_p}(A^*) \in \mathbf{Z}_p$

- $P_{Z_p}(A^* | B^*) \in \mathbf{Q}_p$ is characterized by the following constraint:

$$P_{Z_p}(A^* | B^*) = \frac{P_{Z_p}(A^* \cap B^*)}{P_{Z_p}(B^*)} = \frac{F_{Z_p}(A^* \cap B^*)}{F_{Z_p}(B^*)},$$

where $P_{Z_p}(B^*) \neq 0$, $P_{Z_p}(A^* \cap B^*) = \inf(P_{Z_p}(A^*), P_{Z_p}(B^*))$.

The measure $P_{Z_p}(\cdot)$ runs over the set $\mathbf{Q}_p$ of all *p*-adic numbers (not only integers). Notice that while $\mathbf{Z}_p$ is the ring of *p*-adic integers, $\mathbf{Q}_p$ is the field of *p*-adic numbers.

10.5 States of Knowledge and Strategies of Plasmodium

Using *p*-adic valued fuzziness and probabilities, we can define games of plasmodia. So, in the given topology of attractants, active zones of plasmodia (initial states) can be considered players. Suppose, we have a set of N players, call them $i = 1, ..., N$. Agent i's *knowledge structure* is a function $\mathbf{P}_i$ which assigns to each attractant $\omega \in \Omega^*$ a non-empty subset of Ω^*, so that each thing ω belongs to one or more elements of each $\mathbf{P}_i$, i.e. Ω^* is contained in a union of $\mathbf{P}_i$, but $\mathbf{P}_i$ are not mutually disjoint. Then $\mathbf{P}_i(\omega)$ is called i's knowledge state at the attractant ω. This means that if the actual state is ω, the individual only knows that the actual state is in $\mathbf{P}_i(\omega)$.

We can interpret $\mathbf{P}_i(\omega)$ probabilistically as follows: $\mathbf{P}^i(\omega) = \{\omega' : P^i_{Z_p}(\omega' | \omega) > 0\}$. Evidently that $P^i_{Z_p}(\omega | \omega) > 0$ for all $\omega \in \Omega^*$, therefore for all $\omega \in \Omega^*$, $\omega \in \mathbf{P}_i(\omega)$.

Now we consider the relation $A^* \subseteq \mathbf{P}_i(\omega)$, where $A^* \subseteq \Omega^*$, as the statement that at ω agent i *accepts the performance* A^*:

$$K_i A^* = \{\omega : A^* \subseteq \mathbf{P}_i(\omega)\}.$$

Let B^*_i mean 'Attractants, which can be occupied by agent i'. After several steps, we expect fusions of all protoplasmic tubes so that all attractants are occupying by all agents. Does it mean that we observe a union of B^*_i? No, it does not. We face just the situation that since a time step $t = k$ the sets B^*_i are intersected. Let C^*_i mean 'Attractants accessible for the attractant N_i by protoplasmic tubes'. Assume, $\omega \in B^*_i$ and $\omega' \in C^*_i$. Evidently, $P^i_{Z_p}(\omega' | \omega) > 0$. As a consequence, we assume according to our definitions that each agent i knows ω at ω' and knows ω' at ω, i.e. agent i accepts the performance B^*_i at ω' and i accepts the performance C^*_i at ω.

Let *get Access Set* (i, k) be a set of all attractants such that i knows about them at the given *setTimeStep*(t_0) and *setTimeEnd*(t_k). A strategy of a player i is a mapping $strat_{i,k} : getAccessSet(i, k) \rightarrow \Omega^*$ such that for any history knowledge $getAccessSet(i, k)$ it is true that $strat_{i,k}$ belongs to the set of attractants accessible at k.

In this knowledge structure of slime mould, we can reject the Aumann agreement theorem [93].

Hence, we have examined some basics of bio-inspired game theory on the slime mould in the p-adic valued universe. In the next Chapter, we are going to consider games in the 5-adic universe. These examples of bio-inspired games will concern implementations of Go games on the plasmodium propagation.

Chapter 11
Physarum Go Games and Rough Sets of Payoffs

11.1 Go Games

Go is a game, originated in ancient China, in which two persons play with a Go board and Go stones. In general, two players alternately place black and white stones, on the vacant intersections of a board with a 19×19 grid of lines, to surround the territory. Whoever has more territory at the end of the game is the winner.

As it was shown in [100], the Go game can be simulated by means of a 5-adic *Physarum* machine. Two syllogistic systems implemented as Go games were considered in [96, 100], namely, the Aristotelian syllogistic as well as the performative syllogistic [91, 99, 110, 113]. In the first case, the locations of black and white stones are understood as locations of attractants and repellents, respectively. In the second case, the locations of black stones are understood as locations of attractants occupied by plasmodium of *Physarum polycephalum* and the locations of white stones are understood as locations of attractants occupied by plasmodium of *Badhamia utricularis*. The Aristotelian syllogistic version of the Go game is a coalition game. The performative syllogistic version of the Go game is an antagonistic game.

11.2 Rough Set Based Assessment of Payoffs

In this section, we present the formalism for the rough set version of the Go game implemented on the *Physarum* machine. Such version of the Go game, originally proposed in [122], is an antagonistic game implemented in plasmodia of *Physarum polycephalum* and *Badhamia utricularis*.

Let us take into consideration a board for the Go game, with a 19×19 grid of lines. The set of all intersections of the grid is denoted by I. At the beginning, the fixed numbers of original points of plasmodium of *Physarum polycephalum* as well as plasmodium of *Badhamia utricularis* are randomly deployed on intersections.

© Springer International Publishing AG, part of Springer Nature 2019
A. Schumann and K. Pancerz, *High-Level Models of Unconventional Computations*, Studies in Systems, Decision and Control 159,
https://doi.org/10.1007/978-3-319-91773-3_11

Formally, for the rough set version of the Go game implemented on the *Physarum* machine $\mathscr{P}\mathscr{M}$, it is assumed that the structure of $\mathscr{P}\mathscr{M}$ is a triple:

$$\mathscr{P}\mathscr{M} = (Ph, Ba, Attr),$$

where:

- $Ph = \{ph_1, ph_2, \ldots, ph_k\}$ is the set of original points of the plasmodia of *Physarum polycephalum*.
- $Ba = \{ba_1, ba_2, \ldots, ba_l\}$ is the set of original points of the plasmodia of *Badhamia utricularis*.
- $Attr = \{Attr^t\}_{t=0,1,2,\ldots}$ is the family of the sets of attractants, where $Attr^t = \{attr_1^t, attr_2^t, \ldots, attr_{r_t}^t\}$ is the set of all attractants present at time instant t in $\mathscr{P}\mathscr{M}$.

One can see the assumption that the state of the Go game (i.e. the state of the *Physarum* machine) is observed only at discrete time instants, i.e. $t = 0, 1, 2, \ldots$. The structure of the *Physarum* machine $\mathscr{P}\mathscr{M}$ is changing in time (new attractants can always appear). Positions of original points of the plasmodia as well as attractants are considered in the two-dimensional space of intersections. Hence, each intersection $i \in I$ is identified by two coordinates x and y. This fact will be denoted by $i(x, y)$. For each intersection $i(x, y)$, we can distinguish its adjacent surroundings:

$$Surr(i) = \{i'(x', y') \in I : (x' = x - 1 \vee x' = x + 1) \wedge (x' \geq 1) \wedge (x' \leq 19)$$
$$\wedge (y' = y - 1 \vee y' = y + 1) \wedge (y' \geq 1) \wedge (y' \leq 19)\}.$$

Example 11.1 Let us consider an example of the initial configuration of the Go game shown in Fig. 11.1. One can see two original points of plasmodium of *Physarum polycephalum*, ph_1 and ph_2, interpreted as black stones, as well as two original points of plasmodium of *Badhamia utricularis*, ba_1 and ba_2, interpreted as white stones, all of them deployed on intersections. Formally, we can write that, in the initial configuration, $\mathscr{P}\mathscr{M} = (Ph, Ba, Attr)$, where $Ph = \{ph_1, ph_2\}, Ba = \{ba_1, ba_2\}$, and $Attr^0 = \emptyset$.

During the game, two players alternately place attractants on the vacant intersections of the board to surrounding territory. The first player plays for plasmodium of *Physarum polycephalum*, the second one for plasmodium of *Badhamia utricularis*. The plasmodia look for attractants and propagate protoplasmic veins towards them. The attractants occupied by plasmodium of *Physarum polycephalum* are treated as black stones whereas the attractants occupied by plasmodium of *Badhamia utricularis* are treated as white stones.

Formally, during the game, at given time instant t, we can distinguish three kinds of intersections in the set I^t of all intersections:

- $I_\emptyset^t$—a set of all vacant intersections at t.
- $I_\bullet^t$—a set of all intersections occupied by plasmodia of *Physarum polycephalum* at t (black stones).

- $I_{\circ}^{t}$—a set of all intersections occupied by plasmodia of *Badhamia utricularis* at t (white stones).

One can see that $I^{t} = I_{\emptyset}^{t} \cup I_{\bullet}^{t} \cup I_{\circ}^{t}$, where $I_{\emptyset}^{t}$, $I_{\bullet}^{t}$, and $I_{\circ}^{t}$ are pairwise disjoint.

A set I_{π}^{t} of all intersections occupied by the given plasmodium π (either the plasmodium of *Physarum polycephalum* or the plasmodium of *Badhamia utricularis*) at a given time instant t can be approximated (according to a rough set definition) by surroundings of intersections occupied by these plasmodia. Such approximation will be called surroundings approximation.

The lower surroundings approximation $\underline{Surr}(I_{\pi}^{t})$ of I_{π}^{t} is given by configuration of the Go game.

$$\underline{Surr}(I_{\pi}^{t}) = \{i \in I_{\pi}^{t} : Surr(i) \neq \emptyset \wedge Surr(i) \subseteq I_{\pi}^{t}\},$$

where π is either $\bullet$ or $\circ$. Each intersection $i \in I$ such that $i \in \overline{Surr}(I_{\pi}^{t})$ is called a full generator of the payoff of the player playing for the plasmodium π.

The upper surroundings approximation $\overline{Surr}(I_{\pi}^{t})$ of I_{π}^{t} is given by

$$\overline{Surr}(I_{\pi}^{t}) = \{i \in I_{\pi}^{t} : Surr(i) \cap I_{\pi}^{t} \neq \emptyset\},$$

where π is either $\bullet$ or $\circ$.

The set $BN_{Surr}(I_{\pi}^{t}) = \overline{Surr}(I_{\pi}^{t}) - \underline{Surr}(I_{\pi}^{t})$ is referred to as the boundary region of surroundings approximation of I_{π}^{t} at time instant t. Each intersection $i \in I$ such that $i \in BN_{Surr}(I_{\pi}^{t})$ is called a partial generator of the payoff of the player playing for the plasmodium π.

By replacing the standard set inclusion with the majority set inclusion in the definition of the lower surroundings approximation, according to the VPRSM approach, we obtain the β-lower surroundings approximation:

$$\underline{Surr}^{\beta}(I_{\pi}^{t}) = \{i \in I_{\pi}^{t} : Surr(i) \neq \emptyset \wedge Surr(i) \overset{\beta}{\subseteq} I_{\pi}^{t}\},$$

Each intersection $i \in I$ such that $i \in \underline{Surr}^{\beta}(I_{\pi}^{t})$ is called a full quasi-generator of the payoff of the player playing for the plasmodium π.

On the basis of lower surroundings approximations, we define a measure assessing payoffs of the players. For the first player playing for the *Physarum polycephalum* plasmodia, the payoff measure has the form:

$$\Theta_{\bullet} = card\left(\underline{Surr}(I_{\bullet}^{t})\right).$$

For the second player playing for the *Badhamia utricularis* plasmodia, the payoff measure has the form:

$$\Theta_{\circ} = card\left(\underline{Surr}(I_{\circ}^{t})\right).$$

In a more relaxed case, we have respectively:

$$\Theta_\bullet = card\,(\underline{Surr}^\beta\,(I_\bullet^t)).$$

and

$$\Theta_\circ = card\,(\underline{Surr}^\beta\,(I_\circ^t)).$$

The goal of each player is to maximize its payoff.

Example 11.2 Let us consider an illustrative configuration, shown in Fig. 11.2, of the Go game after several moves, in relation to the initial configuration shown in Fig. 11.1.

In case of a standard definition of rough sets (i.e. the most rigorous case), intersections belonging to lower surroundings approximations $\underline{Surr}(I_\bullet^t)$ and $\underline{Surr}(I_\circ^t)$ of $(I_\bullet^t)$ and $(I_\circ^t)$, respectively, are marked with grey rectangles in Fig. 11.3. The results of the Go game, in case of payoffs defined on the basis of a standard definition of rough sets, are summarized in Table 11.1.

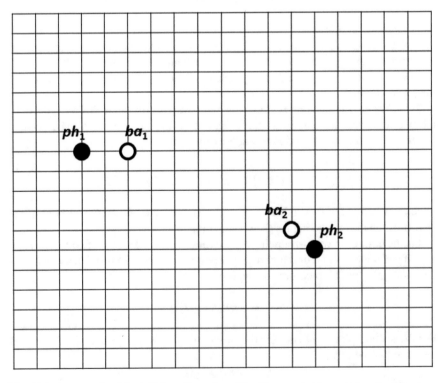

Fig. 11.1 An example of the initial configuration of the Go game implemented on the *Physarum* machine

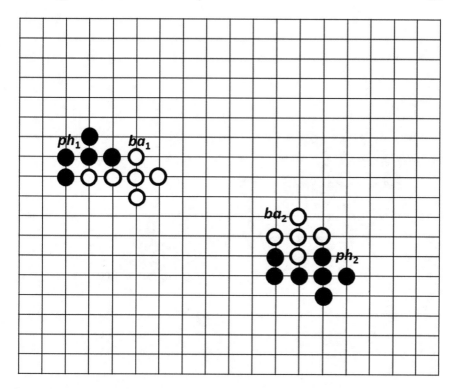

Fig. 11.2 An illustrative configuration of the Go game after several moves

In case of the VPRSM approach (i.e. a more relaxed case), for $\beta = 0.25$, intersections belonging to lower surroundings approximations $\underline{Surr}^{0.25}(I_\bullet^t)$ and $\underline{Surr}^{0.25}(I_\circ^t)$ of ($I_\bullet^t$) and (I_t°), respectively, are marked with grey rectangles in Fig. 11.4. The results of the Go game, in case of payoffs defined on the basis of the VPRSM approach for $\beta = 0.25$, are summarized in Table 11.2.

Let us consider now how we can define strategies in our Go game if we deal with the standard surroundings approximation. A mapping from the intersections belonging to upper surroundings approximations $\overline{Surr}(I_\pi^t)$ at t to the intersections belonging to lower surroundings approximations $\underline{Surr}(I_\pi^{t+k})$ at $t + k$ is said to be a set of rational strategies of π of radius k to win. A mapping from the intersections belonging to β-lower surroundings approximations $\underline{Surr}^\beta(I_\bullet^t)$ at t to the intersections belonging to the union $\underline{Surr}^\gamma(I_\circ^{t+k}) \cup \underline{Surr}^\beta(I_\bullet^{t+k})$ at $t + k$, where $0 < \beta \leq \gamma$ and $card(\underline{Surr}^\beta(I_\bullet^t)) \leq card(\underline{Surr}^\gamma(I_\circ^{t+k}) \cup \underline{Surr}^\beta(I_\bullet^{t+k})) < card(\underline{Surr}(I_\bullet^{t+k}))$, is said to be a set of rational strategies of ∘ (the player playing for *Badhamia utricularis*) of radius k not to lose a game. A mapping from the intersections belonging to β-lower surroundings approximations $\underline{Surr}^\beta(I_\circ^t)$ at t to the intersections belonging to the union $\underline{Surr}^\gamma(I_\bullet^{t+k}) \cup \underline{Surr}^\beta(I_\circ^{t+k})$ at $t + k$, where $0 < \beta \leq \gamma$ and $card(\underline{Surr}^\beta(I_\circ^t)) \leq card(\underline{Surr}^\gamma(I_\bullet^{t+k}) \cup \underline{Surr}^\beta(I_\circ^{t+k})) < card(\underline{Surr}(I_\circ^{t+k}))$, is said to be a set of rational

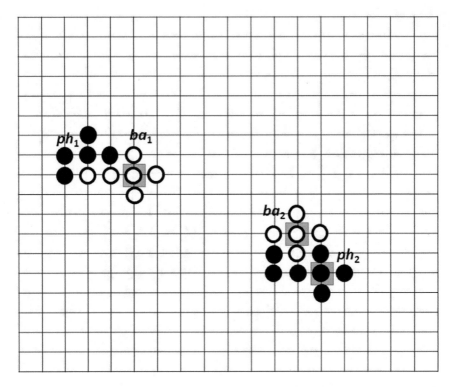

Fig. 11.3 A configuration of the Go game after several moves (payoffs defined on the basis of a standard definition of rough sets)

Table 11.1 The results of the Go game in case of payoffs defined on the basis of a standard definition of rough sets

	The player playing for *Physarum polycephalum* plasmodia	The player playing for *Badhamia utricularis* plasmodia
	Black stones	*White stones*
#Full generators	1	2
Θ_π	1	2
Result		Winner

strategies of ● (the player playing for *Physarum polycephalum*) of radius k not to lose a game.

The agent is rational if (s)he follows one of the rational strategies to win or not to lose in moves. Also, we can define strategies in the Go game if we deal with the VPRSM surroundings approximation. A mapping from the intersections belonging to γ-lower surroundings approximations $\underline{Surr}^\gamma(I_\pi^t)$ at t to the intersections belonging to β-lower surroundings approximations $\underline{Surr}^\beta(I_\pi^{t+k})$ at $t+k$ is said to be a set of rational β-strategies of π of radius k to win if $\gamma < beta$. A

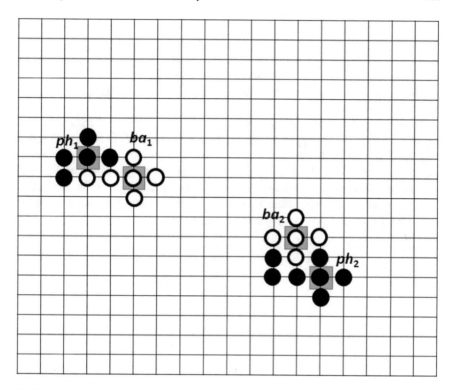

Fig. 11.4 A configuration of the Go game after several moves (payoffs defined on the basis of the VPRSM approach for $\beta = 0.25$)

Table 11.2 The results of the Go game in case of payoffs defined on the basis of the VPRSM approach for $\beta = 0.25$

	The player playing for *Physarum polycephalum* plasmodia	The player playing for *Badhamia utricularis* plasmodia
	Black stones	*White stones*
#Full generators	1	2
#Full quasi-generators	1	0
Θ_π	2	2
Result	No player wins	

mapping from the intersections belonging to γ-lower surroundings approximations $\underline{Surr}^\gamma(I_\bullet^t)$ at t to the intersections belonging to the union $\underline{Surr}^\delta(I_\circ^{t+k}) \cup \underline{Surr}^\gamma(I_\bullet^{t+k})$ at $t + k$, where $0 < \beta \le \gamma$ and $0 < \beta \le \delta$ and $card(\underline{Surr}^\gamma(I_\bullet^t)) \le card(\underline{Surr}^\delta(I_\circ^{t+k}) \cup \underline{Surr}^\gamma(I_\bullet^{t+k})) < card(\underline{Surr}^\beta(I_\bullet^{t+k}))$, is said to be a set of rational β-strategies of $\circ$ (the player playing for *Badhamia utricularis*) of radius k not to lose. A mapping from the intersections belonging to γ-lower surroundings approximations $\underline{Surr}^\gamma(I_\circ^t)$ at t to the intersections belonging to the union $\underline{Surr}^{\delta}(I_{\bullet}^{t+k}) \cup \underline{Surr}^{\gamma}(I_{\circ}^{t+k})$ at $t+k$, where $0 < \beta \leq \gamma$ and $0 < \beta \leq \delta$ and $card(\underline{Surr}^{\gamma}(I_{\circ}^{t})) \leq card(\underline{Surr}^{\delta}(I_{\bullet}^{t+k}) \cup \underline{Surr}^{\gamma}(I_{\circ}^{t+k})) < card(\underline{Surr}^{\beta}(I_{\circ}^{t+k}))$, is said to be a set of rational β-strategies of $\bullet$ (the player playing for *Physarum polycephalum*) of radius k not to lose a game.

The agent is β-rational if (s)he follows one of the rational β-strategies to win or not to lose in moves.

In implementing the Go games on the slime mould, we have distinguished two players (who put attractants or repellents) from two agents of Go games (*Physarum polycephalum* and *Badhamia utricularis*). The players (human beings) program the slime mould behaviour by scattering attractants and repellents intentionally in the right places. Both species of the slime mould compete to follow the human instructions and to calculate some arithmetic functions in the p-adic universe. Thus, we deal with a computer interface in a game-theoretic setting.

Chapter 12
Interfaces in a Game-Theoretic Setting for Controlling the Physarum Motions

Conventionally, the intelligent behaviour of animals is explained by their nervous system that coordinates voluntary and involuntary actions of their bodies and transmits signals between different parts of their bodies, which allow animals to act intentionally and efficiently. There is an approach in artificial intelligence, consisting in building computational models inspired by these nervous systems, that is called *artificial neural networks*.

Nevertheless, *Physarum polycephalum* and *Badhamia utricularis* demonstrate an intelligent behaviour with intentionality and efficiency, although they do not have nervous systems at all. In particular, they demonstrate the ability to memorize and anticipate repeated events [75]. Furthermore, by means of plasmodium behaviour, it is possible to simulate the behaviour of some collectives such as collectives of parasites [111–113]. Thus, the complex intelligent behaviour of plasmodium is still biologically unexplained and shows the limits of our understanding what natural intelligence is.

In this chapter we will show that since we can control the plasmodium motions as a game, the user interfaces for the controllers of plasmodium propagations can have a natural form of game-theoretic setting.

12.1 Software Tool

The *Physarum* software system, called shortly *PhysarumSoft*, is a specialized software tool developed for programming *Physarum* machines and simulating *Physarum* games (see [116]). It was designed and implemented for the Java platform. A general structure of this tool is shown in Fig. 12.1. We can distinguish three main parts of *PhysarumSoft*:

© Springer International Publishing AG, part of Springer Nature 2019
A. Schumann and K. Pancerz, *High-Level Models of Unconventional
Computations*, Studies in Systems, Decision and Control 159,
https://doi.org/10.1007/978-3-319-91773-3_12

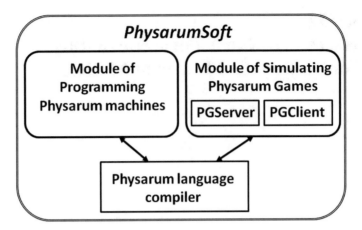

Fig. 12.1 A general structure of *PhysarumSoft*

1. *Physarum* language compiler.
2. Module of programming *Physarum* machines.
3. Module of simulating *Physarum* games.

At the highest level, *PhysarumSoft* uses a ladder diagram, a transition system, a timed transition system, a Petri net, and a tree structure as models of plasmodium propagation in *Physarum* machines. To build such models, a formal object-oriented programming language, called the *Physarum* language, is applied (see Chapter 7). *Physarum* language compiler is used to translate the high-level models into the spatial distribution (configuration) of stimuli (attractants and/or repellents) for *Physarum* machines.

The third part of PhysarumSoft is a module, called the *Physarum* game simulator, designed for simulating games on *Physarum* machines. This module works under the client-server paradigm. A general structure of the *Physarum* game simulator is shown in Fig. 12.2.

The communication between clients and the server is realized through text messages containing statements of the *Physarum* language. The server sends to clients information about the current configuration of the *Physarum* machine (localization of the original points of *Physarum polycephalum* and *Badhamia utricularis*, localization of stimuli, as well as a list of edges, corresponding to veins of plasmodia, between active points) through the XML file.

The server-side application of the *Physarum* game simulator is called *PGServer*. The main window of *PGServer* is shown in Fig. 12.3. In this window, the user can:

- select the port number on which the server listens for connections,
- start and stop the server,
- set the game:

  – a *Physarum* game with strategy based on stimulus placement (see [63]),

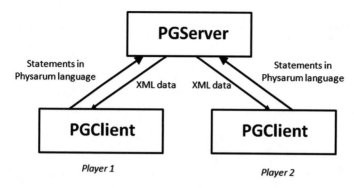

Fig. 12.2 A general structure of the *Physarum* game simulator

Fig. 12.3 The main window of *PGServer*

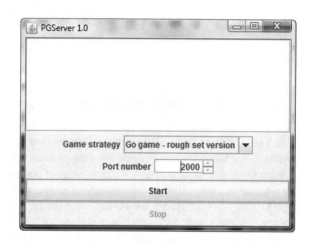

– a *Physarum* game with strategy based on stimulus activation (see [63]),
– a rough set version of the Go game (see Section 11.2),

• shadow information about actions undertaken.

The client-side application of the *Physarum* game simulator is called *PGClient*. The main window of *PGClient* is shown in Fig. 12.4. In case of the Go game, the user can:

• set the server IP address and its port number,
• start the participation in the game,
• put attractants at the vacant intersections of a board,
• monitor the current state of the game as well as the current assessment of payoffs.

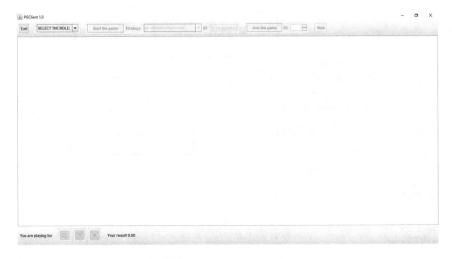

Fig. 12.4 The main window of *PGClient*

12.2   Game-Theoretic Interfaces for Plasmodia

It is known due to the experiments performed by Andrew Adamatzky and Martin Grube that if there are only two agents of the plasmodium game, where the first agent is presented by a usual *Physarum polycephalum* plasmodium and the second agent by its modification called a *Badhamia utricularis* plasmodium, then both of them start to compete with each other. In particular, the *Physarum polycephalum* plasmodium grows faster and could grow into branches of *Badhamia utricularis*, while the *Badhamia utricularis* plasmodium could grow over *Physarum polycephalum* veins. So, we face an interesting form of zero-sum games.

The user interface for this game is designed on the basis of the following rules:

1. *n* agents of *Physarum polycephalum* and *m* agents of *Badhamia utricularis* are generated.
2. Locations of attractants and repellents can be determined.
3. The task is, for example, to reach as many attractants as possible or to construct the longest path consisting of occupied attractants, etc.

The plasmodium game has the form of cycle of Fig. 12.5.

In the *Physarum* game simulator, we have two players:

- the first one plays for the *Physarum polycephalum* plasmodia,
- the second one plays for the *Badhamia utricularis* plasmodia.

In case of a *Physarum* game, locations of the original points of both plasmodia are randomly generated. The players can control motions of plasmodia via attracting or repelling stimuli. There are two strategies which can be defined for the game:

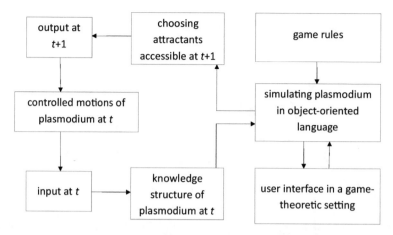

Fig. 12.5 The operative cycle of game-theoretic controller of plasmodium motions

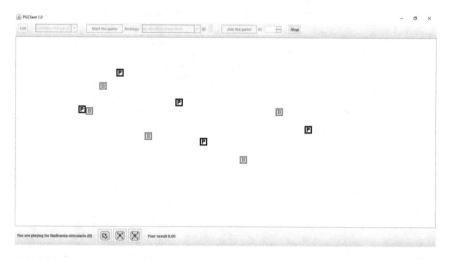

Fig. 12.6 The main window of *PGClient* for the second strategy

1. Locations of attractants and repellents are a priori generated in a random way. During the game, each player can activate one stimulus (attractant or repellent) at each step.
2. Locations of attractants and repellents are determined by the players during the game. At each step, each player can put one stimulus (attractant or repellent) at any location and this stimulus becomes automatically activated.

The client-side main window for the second strategy is shown in Fig. 12.6. At the beginning, the original points of *Physarum polycephalum* and *Badhamia utricularis* are scattered randomly on the plane. During the game, players can place stimuli.

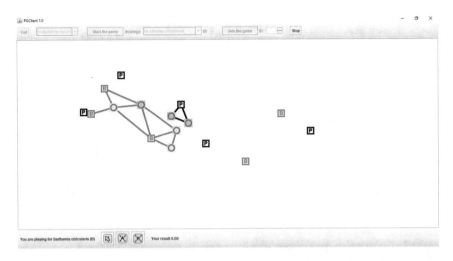

Fig. 12.7 The main window of *PGClient* for the second strategy after several player's movements

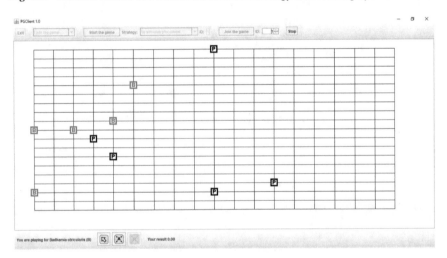

Fig. 12.8 The main window of *PGClient* for the second strategy

New veins of plasmodia are created. The window after several player's movements is shown in Fig. 12.7.

In case of the Go game on the *Physarum* machine, locations of the original points of both plasmodia are randomly generated. The players can control motions of plasmodia via attracting or repelling stimuli placed on the surrounding territory. The client-side main window for the Go game is shown in Fig. 12.8. At the beginning, the original points of *Physarum polycephalum* and *Badhamia utricularis* are scattered randomly on the plane. During the game, players can place stimuli. New veins of

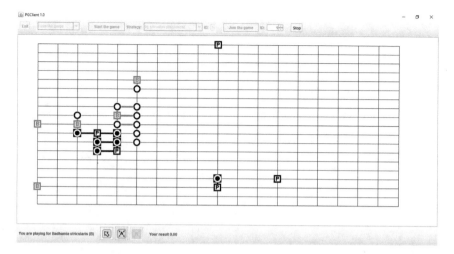

Fig. 12.9 The main window of *PGClient* for the second strategy after several player's movements

plasmodia in surrounding territories are created. The window after several player's movements is shown in Fig. 12.9.

We propose a game-theoretic interface from *Physarum* to humans. Within this interface, we have designed the zero-sum game between plasmodia of *Physarum polycephalum* and *Badhamia utricularis*. The designed game shows aesthetic aspects of plasmodium patterns. The bio-inspired games wake new interests in designing new games and new game platforms.

Chapter 13
Conclusions

In moving, the plasmodium switches its direction or even multiplies in accordance with different bio-signals attracting or repelling its motions, e.g. in accordance with pheromones of bacterial food, which attract the plasmodium, and high salt concentrations, which repel it. So, the plasmodium motions can be controlled by different topologies of attractants and repellents so that the plasmodium can be considered a programmable biological device in the form of a timed transition system, where attractants and repellents determine the set of all plasmodium transitions (Chap. 2). This device is called a *Physarum* machine (Chap. 3). Within this device we can define Petri nets (Chap. 4) and rough sets on transitions (Chap. 5). The universe for the plasmodium propagation is non-well-founded (Chap. 6). Hence, we can obtain a *Physarum* language that formalizes the slime mould behaviour (Chap. 7). Furthermore, we can define p-adic valued logics on these transitions (Chap. 8). These logics describe p-adic arithmetics implemented within *Physarum* machines (Chap. 9). Notice that p-adic logical values of plasmodia transitions allow us to define a knowledge state of plasmodium and its game strategy in occupying attractants as payoffs for the plasmodium (Chap. 10). If we deal with a 5-adic valued logic implemented in the *Physarum* machine, then the slime mould embodies a biological kind of Go game (Chap. 11). We can regard the task of controlling the plasmodium motions as a game and we can design different interfaces in a game-theoretic setting for the controllers of plasmodium transitions by chemical signals (Chap. 12).

Thus, in this book, we have proposed a bio-inspired game theory on plasmodia, i.e. an experimental game theory, where, on the one hand, all basic definitions are verified in the experiments with *Physarum polycephalum* and *Badhamia utricularis* and, on the other hand, all basic algorithms are implemented in the object-oriented language for simulations of plasmodia. Our results allow us to claim that the slime mould can be a model for concurrent games and context-based games. In context-based games, players can move concurrently as well as in concurrent games, but the set of actions

© Springer International Publishing AG, part of Springer Nature 2019
A. Schumann and K. Pancerz, *High-Level Models of Unconventional Computations*, Studies in Systems, Decision and Control 159,
https://doi.org/10.1007/978-3-319-91773-3_13

is ever infinite. In our experiments, we follow the following interpretations of basic entities:

- Attractants as payoffs;
- Attractants occupied by the plasmodium as states of the game;
- Active zones of plasmodium as players;
- Logic gates for behaviours as moves (available actions) for the players;
- Propagation of the plasmodium as the transition table which associates, with a given set of states and a given move of the players, the set of states resulting from that move.

In the *Physarum* game theory we can demonstrate creativity of primitive biological substrates of plasmodia. The point is that plasmodia do not strictly follow spatial algorithms like Kolmogorov-Uspensky machines, but perform many additional actions. So, the plasmodium behaviour can be formalized within strong extensions of spatial algorithms, e.g. within concurrent games or context-based games. We have proposed a game-theoretic visualization of morphological dynamics with nonsymbolic interfaces between living objects and humans. These non-symbolic interfaces are more general than just sonification and have a game-theoretic form.

The object-oriented programming language for *Physarum* behaviour we have constructed, on the one hand, simulates the *Physarum* behaviour and, on the other hand, shows which mathematical tools can be implemented in its behaviour. In particular, we consider *Physarum polycephalum* as simulation model for different context-based games.

Using the object-oriented programming language for *Physarum polycephalum* computing, different logical systems from behavioural logic to illocutionary logic and from context-based games to epistemic logic on non-Archimedean probabilities are implementable. The proposed language can be used for developing programs for *Physarum* machines by the spatial configuration of stimuli. Spatial distribution of stimuli can be identified with a low-level programming language for *Physarum* machines. The created programming language uses the prototype-based approach called also the class-less or instance-based approach. There are inbuilt sets of prototypes corresponding to both the high-level models used to describe behaviour of *Physarum polycephalum*, e.g. ladder diagrams, Petri nets, transition systems and the low-level model (i.e. distribution of stimuli).

The current version of a new software tool, called *PhysarumSoft*, has been developed for:

- Programming *Physarum* machines.
- Simulating *Physarum* games.

This tool was designed for the Java platform.

We can distinguish three main parts of *PhysarumSoft*:

- *Physarum* language compiler.
- Module of programming *Physarum* machines.
- Module of simulating *Physarum* games.

For generating the compiler of the language, the Java Compiler Compiler (JavaCC) tool was used. A compiler translates the high-level code describing the model of *Physarum* machine into the spatial distribution (configuration) of stimuli (attractants, repellents) controlling propagation of protoplasmic veins of the plasmodium.

References

1. Abramsky, S., Jagadeesan, R.: Games and full completeness for multiplicative linear logic. J. Symb. Logic **59**(2), 543–574 (1994)
2. Abramsky, S., Mellies, P.A.: Concurrent games and full completeness. In: Proceedings of the 14th Symposium on Logic in Computer Science, pp. 431–442 (1999)
3. Aczel, A.: Non-Well-Founded Sets. Stanford University Press (1988)
4. Adamatzky, A.: Physarum Machines: Computers from Slime Mould. World Scientific (2010)
5. Adamatzky, A.: Slime mould computing. Int. J. Gen. Syst. **44**(3), 277–278 (2015)
6. Adamatzky, A.: A would-be nervous system made from a slime mold. Artif. Life **21**(1), 73–91 (2015)
7. Adamatzky, A., De Lacy Costello, B., Asai, T.: Reaction-Diffusion Computers. Elsevier (2005)
8. Adamatzky, A., Erokhin, V., Grube, M., Schubert, T., Schumann, A.: Physarum Chip Project: Growing computers from slime mould. Int. J. Unconv. Comput. **8**(4), 319–323 (2012)
9. Adamatzky, A., Ilachinski, A.: Slime mold imitates the united states interstate system. Complex Syst. **21**(1) (2012)
10. Adamatzky, A., de Lacy Costello, B., Dittrich, P., Gorecki, J., Zauner, K.: On logical universality of belousov-zhabotinsky vesicles. Int. J. Gen. Syst. **43**(7), 757–769 (2014)
11. Adamatzky, A., Mayne, R.: Actin automata: Phenomenology and localizations. I. J. Bifurcation and Chaos **25**(2) (2015). https://doi.org/10.1142/S0218127415500303
12. Adamatzky, A., Yang, X., Zhao, Y.: Slime mould imitates transport networks in china. Int. J. Intell. Comput. Cybern. **6**(3), 232–251 (2013). https://doi.org/10.1108/IJICC-02-2013-0005
13. Agerwala, T., Flynn, M.: Comments on capabilities, limitations and 'correctness' of Petri nets. In: Proceedings of the 1st Annual Symposium on Computer Architecture (ISCA'1973), pp. 81–86. Atlanta, USA (1973)
14. Alonso-Sanz, R., Adamatzky, A.: Actin automata with memory. I. J. Bifurc. Chaos **26**(1) (2016)
15. Baaz, M., Fermüller, C., Zach, R.: Systematic construction of natural deduction systems for many-valued logics. In: Proceedings of the 23rd International Symposium on Multiple Valued Logic, pp. 208–213. Sacramento, USA (1993)
16. Bachman, G.: Introduction to p-adic Numbers and Valuation Theory. Academic Press (1964)
17. Baillot, P., Danos, V., Ehrhard, T., Regnier, L.: Believe it or not, AJM's games model is a model of classical linear logic. In: Proceedings of the 12th Annual IEEE Symposium on Logic in Computer Science, LICS '97. IEEE Computer Society, Washington, DC, USA (1997)

18. Ben-Ari, M., Pnueli, A., Manna, Z.: The temporal logic of branching time. Acta Inform. **20**(3), 207–226 (1983)
19. Berzina, T., Dimonte, A., Cifarelli, A., Erokhin, V.: Hybrid slime mould-based system for unconventional computing. Int. J. Gen. Syst. **44**(3), 341–353 (2015)
20. Bouyer, P., Brenguier, R., Markey, N., Ummels, M.: Nash equilibria in concurrent games with büchi objectives. In: Proceedings of the IARCS Annual Conference on Foundations of Software Technology and Theoretical Computer Science (FSTTCS'2011), pp. 375–386. Mumbai, India (2011)
21. Bouyer, P., Brenguier, R., Markey, N., Ummels, M.: Concurrent games with ordered objectives. In: Birkedal, L. (ed.) Foundations of Software Science and Computational Structures. Lecture Notes in Computer Science, vol. 7213, pp. 301–315. Springerg, Berlin, Heidelberg (2012)
22. Brenguier, R.: PRALINE: A tool for computing Nash equilibria in concurrent games. In: Sharygina, N., Veith, H. (eds.) Computer Aided Verification. Lecture Notes in Computer Science, vol. 8044, pp. 890–895. Springer, Berlin, Heidelberg (2013)
23. Caires, L., Cardelli, L.: A spatial logic for concurrency: Part I. Inf. Comput. **186**(2), 194–235 (2003)
24. Caires, L., Cardelli, L.: A spatial logic for concurrency: Part II. Theor. Comput. Sci. **322**(3), 517–565 (2004)
25. Calcagno, C., Cardelli, L., Gordon, A.: Deciding validity in a spatial logic for trees. J. Funct. Prog. **15**, 543–572 (2005)
26. Cardelli, L., Gardner, P., Ghelli, G.: A spatial logic for querying graphs. In: Widmayer, P. (ed.) Proceedings ICALP'02, pp. 597–610. Springer (2002)
27. Copeland, B.J.: Hypercomputation. Minds Mach. **12**(4), 461–502 (2002). https://doi.org/10.1023/A:1021105915386
28. Craig, I.: Object-Oriented Programming Languages: Interpretation. Springer-Verlag, London (2007)
29. Dam, M.: Proof systems for pi-calculus logics. In: de Queiroz, R. (ed.) Logic for Concurrency and Synchronisation, pp. 145–212. Kluwer (2003)
30. Dimonte, A., Berzina, T., Erokhin, V.: Basic transitions of Physarum polycephalum. In: Ganzha, M., Maciaszek, L., Paprzycki, M. (eds.) Proceedings of the 2015 Federated Conference on Computer Science and Information Systems (FedCSIS'2015), pp. 599–606. Lodz, Poland (2015)
31. Dimonte, A., Berzina, T., Pavesi, M., Erokhin, V.: Hysteresis loop and cross-talk of organic memristive devices. Microelectron. J. **45**(11), 1396–1400 (2014). https://doi.org/10.1016/j.mejo.2014.09.009
32. Dourvas, N.I., Sirakoulis, G.C., Adamatzky, A.: Cellular automaton belousov-zhabotinsky model for binary full adder. I. J. Bifurc. Chaos **27**(6), 1–14 (2017)
33. Erokhin, V.: On the learning of stochastic networks of organic memristive devices. IJUC **9**(3–4), 303–310 (2013)
34. Erokhin, V., Howard, G.D., Adamatzky, A.: Organic memristor devices for logic elements with memory. I. J. Bifurc. Chaos **22**(11) (2012). https://doi.org/10.1142/S0218127412502835
35. Fiore, M.P.: A coinduction principle for recursive data types based on bisimulation. In: Proceedings 8th Conference Logic in Computer Science (LICS'93), pp. 110–119 (1993)
36. Gandy, R.: Church's thesis and principles for mechanisms. In: Barwise, J., Keisler, H.J., Kunen K. (eds.) The Kleene Symposium, Studies in Logic and the Foundations of Mathematics, vol. 101, pp. 123–148. Elsevier (1980). https://doi.org/10.1016/S0049-237X(08)71257-6
37. Hennessy, M., Milner, R.: Algebraic laws for nondeterminism and concurrency. JACM **32**(1), 137–161 (1985)
38. Henzinger, T.A., Manna, Z., Pnueli, A.: Timed transition systems. In: de Bakker, J., Huizing, C., de Roever, W., Rozenberg, G. (eds.) Real-Time: Theory in Practice. Lecture Notes in Computer Science, vol. 600, pp. 226–251. Springer, Berlin Heidelberg (1992)
39. Howe, D.J.: Proving congruence of bisimulation in functional programming languages. Inf. and Comp **124** (1996)

40. Jaskuła, B., Szkoła, J., Pancerz, K., Derkacz, A.: Eye-tracking data, complex networks and rough sets: an attempt toward combining them. In: Suzuki, J., Nakano, T., Hess, H. (eds.) Proceedings of the 9th International Conference on Bio-inspired Information and Communications Technologies (BICT'2015), pp. 167–173. New York City, New York, USA (2015)
41. Jones, J., Mayne, R., Adamatzky, A.: Representation of shape mediated by environmental stimuli in physarum polycephalum and a multi-agent model. JPEDS 32(2), 166–184 (2017)
42. Kalogeiton, V.S., Papadopoulos, D.P., Georgilas, I., Sirakoulis, G.C., Adamatzky, A.: Cellular automaton model of crowd evacuation inspired by slime mould. Int. J. Gen. Syst. 44(3), 354–391 (2015)
43. Karpenko, A.: Lukasiewicz's Logics and Prime Numbers. Luniver Press (2006)
44. Keller, R.M.: Formal verification of parallel programs. Commun. ACM 19(7), 371–384 (1976)
45. Khrennikov, A.: Non-Archimedean Analysis: Quantum Paradoxes. Springer-Verlag, Dynamical Systems and Biological Models (1997)
46. Khrennikov, A., Schumann, A.: p-adic physics, non-well-founded reality and unconventional computing. P-Adic Numbers Ultrametric Anal. Appli. 1(4), 297–306 (2009)
47. Khrennikov, A.Y.: p-adic quantum mechanics with p-adic valued functions. J. Math. Phys 32(4), 932–937 (1991)
48. Khrennikov, A.Y.: Interpretations of Probability. VSP Int. Sc. Publishers, Utrecht/Tokyo (1999)
49. Khrennikov, A.Y., Schumann, A.: Logical approach to p-adic probabilities. Bull. Sect. Logic 35(1), 49–57 (2006)
50. Koblitz, N.: p-adic Numbers, p-adic Analysis and Zeta Functions, second edn. Springer-Verlag (1984)
51. Lin, T.Y.: Topological and fuzzy rough sets. In: Słowiński, R. (ed.) Intelligent Decision Support: Handbook of Applications and Advances of the Rough Sets Theory, pp. 287–304. Springer, Netherlands, Dordrecht (1992)
52. Mahler, K.: Introduction to p-adic numbers and their functions, second edn. Cambridge University Press (1981)
53. Markov, A.A., Nagorny, N.M.: The Theory of Algorithms, 1st edn. Springer Publishing Company, Incorporated (2010)
54. Milner, R.: Communicating and Mobile Systems: The π-calculus. Cambridge University Press, Cambridge (1999)
55. Nielsen, M., Rozenberg, G., Thiagarajan, P.S.: Elementary transition systems and refinement. Acta Informatica 29(6), 555–578
56. Nielsen, M., Rozenberg, G., Thiagarajan, P.S.: Elementary transition systems. Theor. Comput. Sci. 96(1), 3–33 (1992)
57. Ntinas, V.G., Vourkas, I., Sirakoulis, G.C., Adamatzky, A.: Oscillation-based slime mould electronic circuit model for maze-solving computations. IEEE Trans. Circuits Syst. 64-I(6), 1552–1563 (2017)
58. Pancerz, K.: Quantitative assessment of ambiguities in plasmodium propagation in terms of complex networks and rough sets. In: Nakano, T., Compagnoni, A. (eds.) Proceedings of the 10th EAI International Conference on Bio-inspired Information and Communications Technologies (BICT'2017), pp. 63–66. Hoboken, USA (2017)
59. Pancerz, K.: Rough sets for trees of executions. In: Ganzha, M., Maciaszek, L., Paprzycki, M. (eds.) Position Papers of the 2017 Federated Conference on Computer Science and Information Systems (FedCSIS'2017), pp. 33–36. Czech Republic, Prague (2017)
60. Pancerz, K., Schumann, A.: Principles of an object-oriented programming language for Physarum polycephalum computing. In: Proceedings of the 10th International Conference on Digital Technologies (DT'2014), pp. 273–280. Zilina, Slovak Republic (2014)
61. Pancerz, K., Schumann, A.: Rough set models of Physarum machines. Int. J. Gen. Syst. 44(3), 314–325 (2015)
62. Pancerz, K., Schumann, A.: Some issues on an object-oriented programming language for Physarum machines. In: Bris, R., Majernik, J., Pancerz, K., Zaitseva, E. (eds.) Applications of Computational Intelligence in Biomedical Technology, Studies in Computational Intelligence, vol. 606, pp. 185–199. Springer International Publishing (2016)

63. Pancerz, K., Schumann, A.: Rough set description of strategy games on Physarum machines. In: Adamatzky, A. (ed.) Advances in Unconventional Computing, Volume 2: Prototypes, Models and Algorithms, Emergence, Complexity and Computation, vol. 23, pp. 615–636. Springer International Publishing (2017)

64. Pawlak, Z.: Rough sets. Int. J. Comput. Inf. Sci. **11**(5), 341–356 (1982)

65. Pawlak, Z.: Rough Sets. Theoretical Aspects of Reasoning about Data. Kluwer Academic Publishers, Dordrecht (1991)

66. Pawlak, Z., Skowron, A.: Rudiments of rough sets. Inf. Sci. **177**, 3–27 (2007)

67. Petri, C.: Kommunikationmit automaten. Schriften des IIM nr. 2, Institut für Instrumentelle Mathematik, Bonn (1962)

68. Petri, R.J.: Eine kleine modification des koch'schen plattenverfahrens. Centralblatt für Bakteriologie und Parasitenkunde **1**, 279–280 (1887)

69. Reisig, W.: Petri Nets. Springer, Berlin (1985)

70. Robinson, A.: Non-Standard Analysis. North-Holland, Studies in Logic and the Foundations of Mathematics (1966)

71. Rutten, J.J.M.M.: Processes as terms: non-well-founded models for bisimulation. Math. Struct. Comp. Sci **2**(3), 257–275 (1992)

72. Rutten, J.J.M.M.: Universal coalgebra: a theory of systems. Theor. Comput. Sci **249**(1), 3–80 (2000)

73. Rutten, J.J.M.M.: Behavioral differential equations: a coinductive calculus of streams, automata, and power series. Theor. Comput. Sci. **308**, 1–53 (2003)

74. Rutten, J.J.M.M.: A coinductive calculus of streams. Math. Struct. Comput. Sci. **15**(1), 93–147 (2005)

75. Saigusa, T., Tero, A., Nakagaki, T., Kuramoto, Y.: Amoebae anticipate periodic events. Phys. Rev. Lett. **100**(1), 018,101 (2008)

76. Schumann, A.: Group theory and p-adic valued models of swarm behaviour. Math. Methods Appl. Sci. https://doi.org/10.1002/mma.4540

77. Schumann, A.: Non-archimedean valued sequent logic. In: Eighth International Symposium on Symbolic and Numeric Algorithms for Scientific Computing (SYNASC'06), pp. 89–92. IEEE Press (2006)

78. Schumann, A.: Non-archimedean valued predicate logic. Bull. Sect. Logic **36**(1–2), 67–78 (2007)

79. Schumann, A.: p-adic multiple-validity and p-adic valued logical calculi. J. Multiple-Valued Logic Soft Comput. **13**(1–2), 29–60 (2007)

80. Schumann, A.: Non-archimedean fuzzy and probability logic. J. Appl. Non-Classical Logic. **18**(1), 29–48 (2008)

81. Schumann, A.: Non-archimedean valued and p-adic valued fuzzy cellular automata. J. Cell. Autom. **3**(4), 337–354 (2008)

82. Schumann, A.: Non-well-founded probabilities and coinductive probability logic. In: Eighth International Symposium on Symbolic and Numeric Algorithms for Scientific Computing (SYNASC'08), pp. 54–57. IEEE Computer Society Press (2008)

83. Schumann, A.: Non-well-founded probabilities on streams. In: Dubois, D., Lubiano, M.A., Prade, H., Gil, M.Á., Grzegorzewski, P., Hryniewicz, O. (eds.) Soft Methods for Handling Variability and Imprecision, pp. 59–65. Springer, Berlin Heidelberg, Berlin, Heidelberg (2008)

84. Schumann, A.: Non-archimedean valued extension of logic $l\pi$ and p-adic valued extension of logic bl. J. Uncertain Syst. **4**(2), 99–115 (2010)

85. Schumann, A.: Proof-theoretic cellular automata as logic of unconventional computing. Int. J. Unconv. Comput. **8**(3), 263–280 (2012)

86. Schumann, A.: Anti-platonic logic and mathematics. Multiple-Valued Logic Soft Comput. **21**(1-2), 53–88 (2013). http://www.oldcitypublishing.com/MVLSC/MVLSCabstracts/MVLSC21.1-2abstracts/MVLSCv21n1-2p53-88Schumann.html

87. Schumann, A.: Unconventional logic for massively parallel reasoning. In: 2013 The 6th International Conference on Human System Interaction (HSI), June 2013, pp. 298–305. IEEE Xplore (2013)

88. Schumann, A.: Non-linear permutation groups on physarum polycephalum. In: The 2014 2nd International Conference on Systems and Informatics (ICSAI 2014), pp. 246–251. IEEE Xplore, Shanghai (2014)

89. Schumann, A.: p-adic valued fuzzyness and experiments with physarum polycephalum. In: 11th International Conference on Fuzzy Systems and Knowledge Discovery (FSKD), pp. 466–472. IEEE Xplore (2014)

90. Schumann, A.: Payoff cellular automata and reflexive games. J. Cell. Autom. **9**(4), 287–313 (2014)

91. Schumann, A.: Physarum syllogistic l-systems. In: Future Computing 2014, The Sixth International Conference on Future Computational Technologies and Applications. ThinkMind (2014)

92. Schumann, A.: Probabilities on streams and reflexive games. Oper. Res. Decis. **24**(1), 71–96 (2014). https://doi.org/10.5277/ord140105

93. Schumann, A.: Reflexive games and non-archimedean probabilities. P-Adic Numbers Ultrametric Anal. Appl. **6**(1), 66–79 (2014)

94. Schumann, A.: *p*-adic valued logical calculi in simulations of the slime mould behaviour. J. Appl. Non-Classical Logics **25**(2), 125–139 (2015). https://doi.org/10.1080/11663081.2015.1049099

95. Schumann, A.: From swarm simulations to swarm intelligence. In: Suzuki, J., Nakano, T., Hess, H. (eds.) Proceedings of the 9th International Conference on Bio-inspired Information and Communications Technologies (BICT'2015), pp. 461–468. New York City, New York, USA (2015)

96. Schumann, A.: Go games on plasmodia of Physarum polycephalum. In: Ganzha, M., Maciaszek, L., Paprzycki, M. (eds.) Proceedings of the 2015 Federated Conference on Computer Science and Information Systems (FedCSIS'2015), pp. 615–626. Lodz, Poland (2015)

97. Schumann, A.: Reversible logic gates on physarum polycephalum. AIP Conf. Proc. **1648**(1), 580,011 (2015)

98. Schumann, A.: Towards context-based concurrent formal theories. Parallel Process. Lett. **25**, 1540,008 (2015)

99. Schumann, A.: Physarum polycephalum syllogistic l-systems and judaic roots of unconventional computing. Stud. Logic Gramm. Rhetoric **44**(1), 181–201 (2016). https://doi.org/10.1515/slgr-2016-0011

100. Schumann, A.: Syllogistic versions of go games on physarum. In: Adamatzky, A. (ed.) Advances in Physarum Machines: Sensing and Computing with Slime Mould, pp. 651–685. Springer International Publishing, Cham (2016)

101. Schumann, A.: Toward a computational model of actin filament networks. In: Proceedings of the 9th International Conference on Bio-inspired Systems and Signal Processing (BIOSIGNALS'2016), pp. 290–297. ScitePress, Rome, Italy (2016)

102. Schumann, A.: Towards slime mould based computer. New Math. Nat. Comput. **12**(2), 97–111 (2016)

103. Schumann, A.: Conventional and unconventional approaches to swarm logic. In: Adamatzky, A. (ed.) Advances in Unconventional Computing: Volume 1: Theory, pp. 711–734. Springer International Publishing, Cham (2017)

104. Schumann, A.: Conventional and unconventional reversible logic gates on physarum polycephalum. Int. J. Parallel Emerg. Distrib. Syst. **32**(2), 218–231 (2017). https://doi.org/10.1080/17445760.2015.1068775

105. Schumann, A.: On arithmetic functions in actin filament networks. In: Nakano, T., Compagnoni, A. (eds.) Proceedings of the 10th EAI International Conference on Bio-inspired Information and Communications Technologies (BICT'2017). Hoboken, USA (2017)

106. Schumann, A.: P-adic valued models of swarm behaviour. AIP Conf. Proc. **1863**(1), 360,009 (2017)

107. Schumann, A., Adamatzky, A.: Towards semantical model of reaction-diffusion computing. Kybernetes **38**(9), 1518–1531 (2009)

108. Schumann, A., Adamatzky, A.: Physarum spatial logic. New Math. Nat. Comput. **7**(3), 483–498 (2011)
109. Schumann, A., Adamatzky, A.: The double-slit experiment with physarum polycephalum and p-adic valued probabilities and fuzziness. Int. J. Gen. Syst. **44**(3), 392–408 (2015)
110. Schumann, A., Adamatzky, A.: Physarum polycephalum diagrams for syllogistic systems. IfCoLog J. Logics Appl. **2**(1), 35–68 (2015)
111. Schumann, A., Akimova, L.: Simulating of schistosomatidae (trematoda: Digenea) behavior by physarum spatial logic. In: Ganzha, M., Maciaszek, L., Paprzycki, M. (eds.) Proceedings of the 2013 Federated Conference on Computer Science and Information Systems (FedC-SIS'2013), pp. 225–230. Krakow, Poland (2013)
112. Schumann, A., Akimova, L.: Process calculus and illocutionary logic for analyzing the behavior of schistosomatidae (trematoda: Digenea). In: Pancerz, K., Zaitseva, E. (eds.) Computational Intelligence, Medicine and Biology: Selected Links, pp. 81–101. Springer International Publishing, Cham (2015)
113. Schumann, A., Akimova, L.: Syllogistic system for the propagation of parasites. the case of schistosomatidae (trematoda: Digenea). Stud. Logic Grammar Rhetoric **40**(53), 303–319 (2015)
114. Schumann, A., Fris, V.: Swarm intelligence among humans—the case of alcoholics. In: Proceedings of the 10th International Conference on Bio-inspired Systems and Signal Processing (BIOSIGNALS'2017), pp. 17–25. ScitePress, Porto, Portugal (2017)
115. Schumann, A., Kuznetsov, A.V.: Talmudic foundations of mathematics. In: Nakano, T., Compagnoni, A. (eds.) Proceedings of the 10th EAI International Conference on Bio-inspired Information and Communications Technologies (BICT'2017). Hoboken, USA (2017)
116. Schumann, A., Pancerz: PhysarumSoft—a software tool for programming Physarum machines and simulating Physarum games. In: Ganzha, M., Maciaszek, L., Paprzycki, M. (eds.) Proceedings of the 2015 Federated Conference on Computer Science and Information Systems (FedCSIS'2015), pp. 607–614. Lodz, Poland (2015)
117. Schumann, A., Pancerz, K.: Towards an object-oriented programming language for Physarum polycephalum computing. In: Szczuka, M., Czaja, L., Kacprzak, M. (eds.) Proceedings of the Workshop on Concurrency, Specification and Programming (CS&P'2013), pp. 389–397. Warsaw, Poland (2013)
118. Schumann, A., Pancerz, K.: Timed transition system models for programming Physarum machines: Extended abstract. In: Popova-Zeugmann, L. (ed.) Proceedings of the Workshop on Concurrency, Specification and Programming (CS&P'2014), pp. 180–183. Chemnitz, Germany (2014)
119. Schumann, A., Pancerz, K.: Towards an object-oriented programming language for Physarum polycephalum computing: A Petri net model approach. Fundam. Inf. **133**(2–3), 271–285 (2014)
120. Schumann, A., Pancerz, K.: Interfaces in a game-theoretic setting for controlling the plasmodium motions. In: Proceedings of the 8th International Conference on Bio-inspired Systems and Signal Processing (BIOSIGNALS'2015), pp. 338–343. ScitePress, Lisbon, Portugal (2015)
121. Schumann, A., Pancerz, K.: Petri net models of simple rule-based systems for programming Physarum machines: Extended abstract. In: Suraj, Z., Czaja, L. (eds.) Proceedings of the 24th International Workshop on Concurrency, Specification and Programming (CS&P'2015), vol. 2, pp. 155–160. Rzeszow, Poland (2015)
122. Schumann, A., Pancerz, K.: A rough set version of the Go game on Physarum machines. In: Suzuki, J., Nakano, T., Hess, H. (eds.) Proceedings of the 9th International Conference on Bio-inspired Information and Communications Technologies (BICT'2015), pp. 446–452. New York City, New York, USA (2015)
123. Schumann, A., Pancerz, K.: Roughness in timed transition systems modeling propagation of plasmodium. In: Ciucci, D., Wang, G., Mitra, S., Wu, W.Z. (eds.) Rough Sets and Knowledge Technology, Lecture Notes in Artificial Intelligence, vol. 9436, pp. 482–491. Springer International Publishing (2015)

124. Schumann, A., Pancerz, K.: Logics for physarum chips. Stud. Hum. **5**(1), 16–30 (2016). https://doi.org/10.1515/sh-2016-0002
125. Schumann, A., Pancerz, K.: p-adic computation with Physarum. In: Adamatzky, A. (ed.) Advances in Physarum Machines: Sensing and Computing with Slime Mould, Emergence, Complexity and Computation, vol. 21, pp. 619–649. Springer International Publishing (2016)
126. Schumann, A., Pancerz, K.: Physarumsoft: An update based on rough set theory. AIP Conf. Proc. **1863**(1), 360,005 (2017)
127. Schumann, A., Pancerz, K., Adamatzky, A., Grube, M.: Bio-inspired game theory: The case of Physarum polycephalum. In: Suzuki, J., Nakano, T. (eds.) Proceedings of the 8th International Conference on Bio-inspired Information and Communications Technologies (BICT'2014), pp. 9–16. Boston, Massachusetts, USA (2014)
128. Schumann, A., Pancerz, K., Adamatzky, A., Grube, M.: Context-based games and Physarum polycephalum as simulation model. In: Proceedings of the Workshop on Unconventional Computation in Europe. London, UK (2014)
129. Schumann, A., Pancerz, K., Jones, J.: Towards logic circuits based on Physarum polycephalum machines: The ladder diagram approach. In: Cliquet Jr., A., Plantier, G., Schultz, T., Fred, A., Gamboa, H. (eds.) Proceedings of the International Conference on Biomedical Electronics and Devices (BIODEVICES'2014), pp. 165–170. Angers, France (2014)
130. Schumann, A., Pancerz, K., Szelc, A.: The swarm computing approach to business intelligence. Stud. Hum. **4**(3), 41–50 (2015). https://doi.org/10.1515/sh-2015-0019
131. Schumann, A., Szelc, A.: Towards new probabilistic assumptions in business intelligence. Stud. Hum. **3**(4), 11–21 (2014). https://doi.org/10.1515/sh-2015-0003
132. Schumann, A., Woleński, J.: Decisions involving databases, fuzzy databases and codatabases. Oper. Res. Decis. **25**(3), 59–72 (2015). https://doi.org/10.5277/ord150304
133. Searle, J.R., Vanderveken, D.: Foundations of Illocutionary Logic. Cambridge University Press, Cambridge (1984)
134. Semertzidou, C., Dourvas, N.I., Tsompanas, M.I., Adamatzky, A., Sirakoulis, G.C.: Introducing chemotaxis to a mobile robot. In: Artificial Intelligence Applications and Innovations - 12th IFIP WG 12.5 International Conference and Workshops, AIAI 2016, Thessaloniki, Greece, September 16-18, 2016, Proceedings, pp. 396–404 (2016)
135. Shirakawa, T., p. Gunji, Y., Miyake, Y.: An associative learning experiment using the plasmodium of physarum polycephalum. Nano Commun. Netw. **2**, 99–105 (2011)
136. Shirakawa, T., Sato, H., Ishiguro, S.: Constrcution of living cellular automata using the physarum polycephalum. Int. J. Gen. Syst. **44**, 292–304 (2015)
137. Shirakawa, T., Yokoyama, K., Yamachiyo, M., Y-p, G., Miyake, Y.: Multi-scaled adaptability in motility and pattern formation of the physarum plasmodium **4**, 131–138 (2012)
138. Siccardi, S., Adamatzky, A.: Actin quantum automata: Communication and computation in molecular networks. Nano Comm. Netw. **6**(1), 15–27 (2015). https://doi.org/10.1016/j.nancom.2015.01.002
139. Siccardi, S., Adamatzky, A.: Communication and computation in molecular networks. Actin quantum automata. Nano Commun. Netw. **6**(1), 15–27 (2015)
140. Siccardi, S., Adamatzky, A.: Logical gates implemented by solitons at the junctions between one-dimensional lattices. Int. J. Bifurc. Chaos **26**(6), 1650,107 (2016)
141. Siccardi, S., Adamatzky, A.: Quantum actin automata and three-valued logics. IEEE J. Emerg. Sel. Topics Circuits Syst. **6**(1), 53–61 (2016)
142. Siccardi, S., Tuszynski, J.A., Adamatzky, A.: Boolean gates on actin filaments. Phys. Lett. A **380**(1), 88–97 (2016)
143. Taylor, B., Adamatzky, A., Greenman, J., Ieropoulos, I.: Physarum polycephalum: Towards a biological controller. Biosystems **127**, 42–46 (2015)
144. Tsompanas, M.I., Adamatzky, A., Ieropoulos, I., Phillips, N., Sirakoulis, G.C., Greenman, J.: Cellular non-linear network model of microbial fuel cell. Biosystems **156**, 53–62 (2017)
145. Westendorf, C., Gruber, C., Grube, M.: Quantitative comparison of plasmodial networks of different slime molds. In: Suzuki, J., Nakano, T., Hess, H. (eds.) Proceedings of the 9th International Conference on Bio-inspired Information and Communications Technologies (BICT'2015), pp. 611–612. New York City, New York, USA (2015)

146. Whiting, J.G.H., de Lacy Costello, B., Adamatzky, A.: Sensory fusion in physarum poly-cephalum and implementing multi-sensory functional computation. Biosystems **119**, 45–52 (2014)
147. Whiting, J.G.H., de Lacy Costello, B., Adamatzky, A.: Transfer function of protoplasmic tubes of physarum polycephalum. Biosystems **128**, 48–51 (2015)
148. Winskel, G., Nielsen, M.: Models for concurrency. In: Abramsky, S., Gabbay, D.M., Maibaum, T.S.E. (eds.) Handbook of Logic in Computer Science, vol. 4, pp. 1–148. Oxford University Press (1995)
149. Yao, Y., Zhou, B.: Naive bayesian rough sets. In: Yu, J., Greco, S., Lingras, P., Wang, G., Skowron, A. (eds.) Rough Set and Knowledge Technology: 5th International Conference, RSKT 2010, Beijing, China, October 15–17, 2010. Proceedings, pp. 719–726. Springer Berlin Heidelberg, Berlin, Heidelberg (2010)
150. Yao, Y.Y., Wong, S.K.M., Lin, T.Y.: A review of rough set models. In: Rough Sets and Data Mining: Analysis of Imprecise Data, pp. 47–75. Kluwer Academic Publishers, Dordrecht (1997)
151. Ziarko, W.: Variable precision rough set model. J. Comput. Syst. Sci. **46**(1), 39–59 (1993)

Index

© Springer International Publishing AG, part of Springer Nature 2019
A. Schumann and K. Pancerz, *High-Level Models of Unconventional
Computations*, Studies in Systems, Decision and Control 159,
https://doi.org/10.1007/978-3-319-91773-3

Printed in the United States
By Bookmasters